Real Estate Analysis

Real Estate Analysis: A Toolkit for Property Analysts presents economic and financial models, applications and insights, packaged as a toolkit for analysts and other participants in commercial and residential real estate markets.

Participants in property markets – analysts, brokers, commentators as well as investors and tenants – move seamlessly across a range of physical and financial markets. They employ models that illuminate market activity: the tools of supply and demand to explain rental trends and to forecast vacancy rates and construction cycles; forecasts of macro-economists foreshadow shoppers' spending behaviour in shopping malls and the growth in demand for office space; capital market arithmetic to apply discount and capitalisation rates. Currently these topics are often scattered through textbooks. This book brings these tools together and situates them in a real estate market context.

Topics addressed include:

- The interaction of markets – capital, space and physical assets
- Debt, the cost of capital and investment hurdle rates
- Real options – valuing lease contracts and land
- Risk – what counts, what doesn't (systemic and non-systemic risk)
- Discounted rates and capitalisation rates – interpreting spreads to sovereign bond yields
- Externalities – why do markets "fail"; what are the "solutions"?
- Property rights – different rules, different outcomes
- Exploitation for natural resources (exhaustible, renewable) – how does discounted cash flow analysis (DCF) fit in?
- Cost-benefit analysis – the analytics of compensation payments
- Forecasting – purpose and process

The *foundations* and the *scaffolding* that underpin and support real estate market analysis are the focus of this book. Its purpose is to complement, sometimes augment, the subject matter of real estate training programs. The prospective audience includes curious professionals and researchers, seeking perspectives that extend standard class-room fare.

David Rees is an independent consultant. With a background in finance, real estate and investment strategy he has held positions as

- Regional Director of Research at JLL, an international real estate broker and manager
- Head of Research at Mirvac, a diversified publicly listed REIT
- Head of Research at Commonwealth Bank, Australia's largest commercial bank

He is based in Sydney, Australia.

Real Estate Analysis

A Toolkit for Property Analysts

David Rees

LONDON AND NEW YORK

Designed cover image: © Daniel Berehulak/Staff/Getty Images

First published 2024
by Routledge
4 Park Square, Milton Park, Abingdon, Oxon OX14 4RN

and by Routledge
605 Third Avenue, New York, NY 10158

Routledge is an imprint of the Taylor & Francis Group, an informa business

British Library Cataloguing-in-Publication Data
A catalogue record for this book is available from the British Library

ISBN: 978-0-367-63026-3 (hbk)
ISBN: 978-0-367-63021-8 (pbk)
ISBN: 978-1-003-11193-1 (ebk)

DOI: 10.1201/9781003111931

Typeset in Times New Roman
by codeMantra

Contents

Foreword
A little learning

Mark Twain is alleged to have said,

> What gets us into trouble is not what we don't know. It's what we know for sure that just ain't so.

Property market analysts are intellectually promiscuous. We stroll the aisles of a cerebral supermarket, selecting and discarding freely from a dazzling display of tools and financial theories that claim to illuminate property market activity.

We invoke the tools of supply and demand to explain rental trends, forecast vacancy rates and anticipate construction cycles. The hog-corn cycle helps us to explain time lags in office market construction responses. We borrow from the forecasts of macro-economists to foreshadow shoppers' spending behaviour in retail malls and borrow from demographers to project the growth in demand for office and logistics space. To ascertain real estate market values we harness capital market arithmetic to derive discount and (more problematically) capitalisation rates. For long-term valuation benchmarks we often appeal to the "risk-free" sovereign bond rate with some adjustment for a variety of risks, nodding an acknowledgement to the Capital Asset Pricing Model (CAPM) and portfolio theory.

As we perform these tasks we rummage around in our intellectual toolbox for the most convenient implement, model, insight or concept, sometimes recalling a half-forgotten undergraduate tutorial or textbook.

The task of integrating formal economic concepts and financial models into real estate analysis is complex because property markets are a favourite target for taxes and regulators, which disturb the assumption of frictionless markets that underlie most financial models and economic theories.

However, like a lighthouse on a stormy night, these market interventions enhance, they do not reduce, the insights of economic models and financial principles. Real estate markets are typically hedged about with location-specific rules and regulations, taxes and zoning restrictions that can limit, stimulate or divert activity, and which are often subject to sharp changes in government or urban planning policy. These changes, if they are unanticipated, or unforeseeable, can arbitrarily create winners and losers, raising profound questions of equity (or "fairness") as well as economic efficiency and, inevitably, opportunities for rent-seeking. Predictably, potential losers will often oppose shifting the goalposts, whatever the broader public interest, raising valid questions of compensation and reparation. Further, real estate assets are heterogenous: every asset is unique; markets are often opaque and illiquid. Unlike financial markets, performance is often measured by appraised, not market, valuations, stimulating debate about data validity, risk and performance measurement.

While the application of economic and financial models to real estate markets has many challenges, the rewards can be equally significant.

Real estate analysis, and the cognate disciplines of economics, law and finance, has attracted the attention of more than its fair share of eminent economists – Nobel Memorial Prize in Economic Sciences laureats; Douglass North, Ronald Coase, James Meade, James Buchanan, William Sharpe, Bengt Holmstrom, Franco Modigliani and Merton Miller as well as John Hicks spring readily to mind. Then there are the specific property rights-related insights of Gordon Tulloch, Harold Demsetz, Yoram Barzel, Steven Cheung and Armen Alchian. And, of course, a long list of classical writers – John Stuart Mill, Thomas Malthus, Alfred Marshall, Karl Marx, David Ricardo – grappled with problems of demography, site valuation, public goods, ownership arrangements and the exploitation of exhaustible resources, all so resonant with our own modern concerns. Most of these authorities receive their five minutes of fame in these pages. They deserve to be recorded… better, they deserve to be explored.

Nor are real estate market insights confined to academic journals. Anyone who scans the United States' mid-West fly-over states from a window seat at 25,000 feet – or indeed the top-down topography of England's home counties or parts of Eastern Europe from a similar vantage point – will see them from a new perspective after reading *Measuring America*, by Andro Linklater.[1]

I hope some of these chapters will offer an additional insight or, better, raise a niggling doubt on topics such as the calculation of compensation payments to expropriated homeowners, the ownership drivers of urban consolidation policies, the assessment of vacant land values, policies to better manage the environment and the contribution of "cheap" debt to achieve the required hurdle rate for a development project.

Some chapters are discursive. Others are more technical – familiarity with a micro-economics 101 text is assumed – although I have tried, not always successfully, to restrict mathematical stuff to Appendices. I appeal to intuition wherever possible; so, do skip the technical stuff if you wish – it's the insights that count.

Information is malleable; but wisdom is not a linear process, sliced into convenient 13-week semester segments. It is a long walk along a winding road; we glance into partially open doors, arrested sometimes by flickering shadows that lurk on the blind. I hope that this book is an incentive to stop, turn aside and cautiously explore. That's why I offer no apology for the frequent footnotes to academic literature. Every good guidebook identifies the classic taverns and warns against the tourist traps along the way.

It's the journey, not the destination, that matters.

The *foundations* and the *scaffolding* that underpin and support real estate market analysis are the focus of this book. The chapter of real estate market forecasting concludes with the observation that a successful forecasting program may leave its customers with *less confidence* about the future than they had when they started – but with *more knowledge*.

And that, perhaps, is all that Mark Twain was really trying to say.

Note

1 Linklater, A. *Measuring America* (HarperCollins, 2002).

1 Space, time and assets

The three markets that drive real estate

Property assets sit at the intersection of three distinct but interrelated markets – the market for space (the rental or owner-occupier market), the market for time (the capital market) and the tangible (or physical) property market – the market for assets.

The interaction between these three markets dictates financial performance, market cycles and physical activity. For example, a change in the demand for space drives the rental market and therefore yields (in the capital market) leading to a change in capital values; this in turn drives construction activity in the physical asset market. Property market cycles typically arise from the interaction of these three markets. The three markets themselves reflect the impact of broader social and economic drivers – economic growth, demographic shifts, global and local financial trends, market regulation as well as less easily identified factors such as investor confidence and risk aversion. All these factors express themselves, however, through the three real estate markets.

The three-market model described here offers insights into property market dynamics and a range of concepts such as "yield accretive" investment. Apparently contradictory trends, such as tightening yields at a time when rents are falling, are more readily understood when the three markets are examined as an integrated whole.

Introduction: Introducing the cast

As property market participants – analysts, investors, landlords, tenants and financiers – we move seamlessly and frequently across a range of markets and metrics.

Frequently we view these markets in a largely disconnected fashion and rummage in the analyst's toolbox for the appropriate tool – cash flow analysis, yields, discount rates, rent-free incentives, site values, construction cost estimates and macroeconomic forecasts – as occasion demands.

In his Presidential Address to the American Real Estate and Urban Economics Association in 1990, Jeffrey Fisher offered a simple graphical model that integrated the markets for space and capital.[2] A subsequent paper by DiPasquale and

DOI: 10.1201/9781003111931-1

Wheaton (D–W) amplified this approach.[3] This chapter presents a modified version of the D–W model.

The model will then be used to interpret some of the trends we observe in commercial property markets. For example, the model provides insights into "yield accretive" investments in existing assets. The model helps us understand, for example, the "anomaly" of tightening yields coinciding with falling rents as well as the rationale for investors to pay "above market" for assets.

Defining the three markets

First, *the market for space* is the province of landlords, leasing agents and tenants. The physical demand for commercial space is often analysed as a response (sometimes lagging or leading) to macro-economic drivers such as Gross Domestic Product (GDP) growth, or other economic variables such as employment, consumer spending or investment and wages growth. The physical supply of space can often be regarded as fixed in the short run, though in the long run lagged supply can display high elasticity in response to changes, or expected changes, in rental levels or capitalisation rates.

Second, the *capital market* puts a price on *time*. This market is the province of bankers, investors and property valuers. The yield (or the related capitalisation rate) converts the future stream of rental income to today's capital value.[4] The concept of "Years' Purchase" sometimes still used by real estate analysts is an explicit statement that yields and discount rates put a financial value on time.

Third, the *physical market* for assets determines *construction, demolition and refurbishment activity* which in turn influences rents and market values in the *market for space*. This market is the province of architects, developers, engineers, quantity surveyors, investors and sales agents. Construction costs and land values determine the volume and pace of construction activity but are limited by financial, technological or regulatory constraints, which ultimately determine the area, the quality, the end-use and the height of every building.[5]

Identifying and forecasting the external drivers that cause shifts in these three markets, then analysing and disentangling the inter-relationships between these markets is a core responsibility of real estate investment analysts.

Putting the three markets together

Each of the three markets is presented individually in graphical form. They are then linked as an integrated whole.

The market for space: Consider a hypothetical office building, say, 173 Pitt Street in the City of London (Figure 1.1). This building has a Net Lettable Area (NLA) of 40,000 sqm. The supply of space (S) at this address is fixed in the short run, represented by the vertical supply line, S. Therefore, fluctuations in rent are determined entirely by shifts in demand for space at this address (DD). The slope of DD is likely to vary between markets and through time. For example, when vacancy is low and demand for office space is high, DD will slope steeply. Since the City of London is a large and competitive market the slope of the demand line,

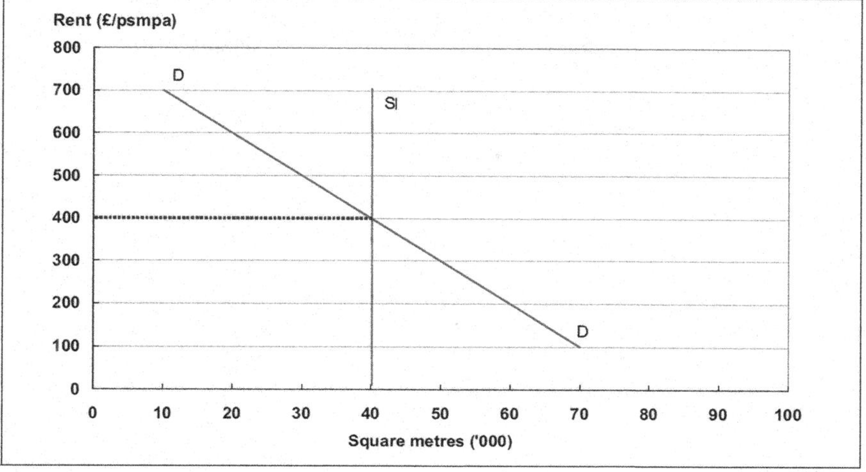

Figure 1.1 The Market for Space: 173 Pitt Street

DD, is likely to be relatively flat, even horizontal if market vacancy rates are high. A horizontal DD line implies that the landlord of 173 Pitt Street (and the tenant) is a price-taker in the market for rented space in the City of London at this time. A price-taker has no ability to influence the rent, which is determined by broad market supply and demand forces.

As shown in Figure 1.1, at an average rent of £400 per square meter per annum (psmpa) total income for the property is £16 million per annum (£400 psmpa × 40,000 sqm).

The capital market: The capital market converts the current and future rental stream from 173 Pitt Street into today's capital value (Figure 1.2). The conversion rate (the yield or capitalisation rate) is assumed to be 8.0 percent in Figure 1.2.

$$£400.00/0.08 = £5000.00 \quad (1.1)$$

Implicit in Eq (1.1) is the assumption that the rent will never change and will be paid in perpetuity. Also note another important qualification. Although Eq (1.1) is a standard feature of many texts[6] it is subject to an important and often overlooked assumption. Eq (1.1) is strictly accurate only if the rent is paid annually *in arrears*. If the rent on 173 Pitt Street is paid *in advance*, as is common in real estate contracts, Eq (1.1) understates the actual value of 173 Pitt Street by the amount of the first annual rent instalment, £400.00. Alternatively, the yield is overstated:

$$£400/£5,400 = 0.074 = 7.4\%$$

Short-run considerations such as variations in expected rental growth, lease expiries and uncertainty regarding planning regulations for example may lead to temporary

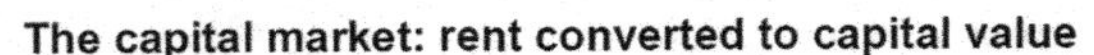

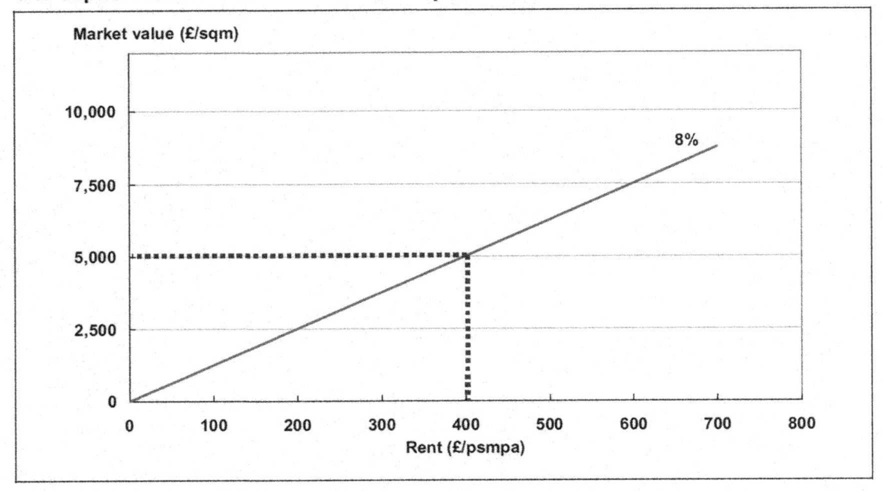

Figure 1.2 The Capital Market: Rent Converted to Capital Value

fluctuations and regional or asset-specific variations in capitalisation rates. In the long run, however, the capitalisation rate is established by general conditions in financial and rental markets and the specific characteristics of each particular market, in this case, the City of London, and the individual asset, in this case, 173 Pitt Street.[7]

We assume now that the rent on 173 Pitt Street is paid annually in arrears. At an 8.0 percent capitalisation rate, £400 psmpa converts to a capital value of £5,000.00 psm. On this basis, 173 Pitt Street has a market value of £5,000 psm × 40,000 sqm = £200 million. When the capitalisation rate falls, the line in Figure 1.2 rotates anti-clockwise. A 6.0 percent capitalisation rate (with the rent unchanged at £400.00 psmpa) converts to a value of

$$£400.00/0.06 = £6.666.67 \text{ psm}$$

The physical market for assets: Why was 173 Pitt Street constructed as a 40,000 sqm NLA building, rather than, say 30,000 sqm or 50,000 sqm?

Figure 1.3 answers this question with a conventional micro-economic diagram that defines the optimisation problem that confronts the developer of 173 Pitt Street.

As the NLA of the building increases it is assumed that the average construction cost per sqm (shown by the average cost (AC) line in Figure 1.3) rises steadily. At an NLA of 40,000 sqm the construction cost per sqm = £2,500, so the total construction cost is

$$£2,500 \times 40,000 \text{ sqm} = £100 \text{ million}$$

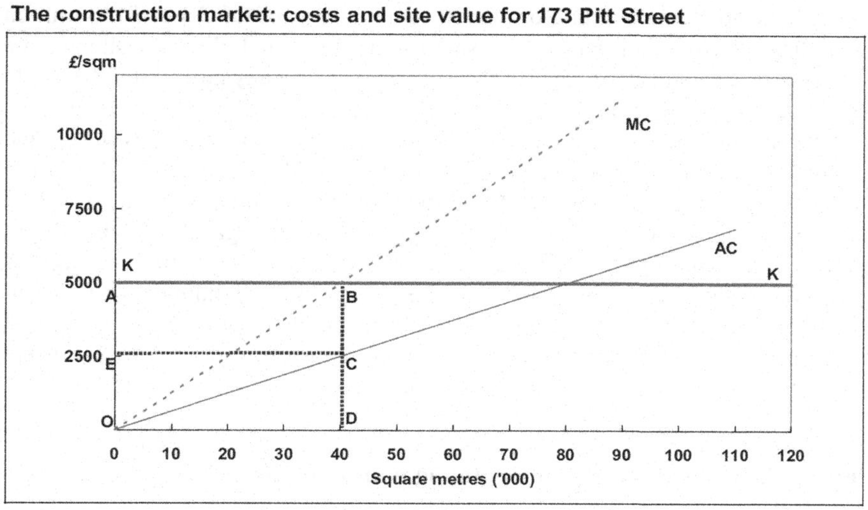

The construction market: costs and site value for 173 Pitt Street

Figure 1.3 The Construction Market: Costs and Site Value for 173 Pitt Street

This is area OECD.

The marginal cost (MC) measures the addition to the total cost of each additional square meter of NLA. For example, an increase in the NLA from 40,000 sqm to 40,001 sqm adds £5,000 to Total Cost which now rises to £100.005 million.

Note that area OECD = area OBD, the area under the MC line.

We know (Figure 1.2) that 173 Pitt Street has a market value of £200 million (OABD). Therefore, the developer's net profit = £200 million − £100 million = £100 million = EABC = OAB.

(We describe this as the developer's "net profit" because we are adopting the economists' convention that the developer's required or "normal" profit is a cost of doing business and is therefore included as a cost in the AC curve.)

Given the rent (£400 psmpa) and capitalisation rate (8.0 percent), £100 million is the maximum net profit achievable from developing this site. As confirmation, the value per sqm = £5000.00 (Figure 1.2) and the marginal cost (MC) line in Figure 1.3 intersects at B at an NLA of 40,000 sqm.

In selecting the optimal or profit maximising NLA of 173 Pitt Street each additional square meter adds £5000.00 to the value of 173 Pitt Street while the addition to cost is less than £5000.00 for any NLA < 40,000 sqm. Therefore, net profit rises to a total NLA of 40,000 sqm. Beyond this MC > £5000 sqm, therefore each additional square meter subtracts from profit.

If the vacant site at 173 Pitt Street had been auctioned prior to development, the winning bid for the site would have been £100 million and the successful bidder would have erected a building of 40,000 sqm at a cost of OECD or £100 million (including developer's profit) on the site. Given the value bid for the site and the construction cost and rental level as defined in Figure 1.3 any other outcome would have resulted in a financial loss.[8]

"Diminishing returns" are here represented as rising average cost (AC) and marginal cost (MC) as in Figure 1.3. The binding limitation on NLA for 173 Pitt Street may arise from engineering considerations (as defined in Figure 1.3) or a regulatory regime such as statutory height limitations; but regardless of the source, there is a finite limit to the dimensions of 173 Pitt Street. While the limiting factor may vary for individual cases, even the largest office building is of finite dimensions. Therefore, a constraint exists. The specific source of the constraint on building size, whether financial, technical or regulatory, is irrelevant to our current purposes.

Putting the model to work – interpreting market trends

Armed with these three diagrams we shall now consider how they all work together. In Figure 1.4 the three markets (Figures 1.1–1.3) now occupy individual quadrants. The fourth (south-easterly) quadrant aligns the area of 173 Pitt Street (40,000 sqm) in Figure 1.1 with the equilibrium construction cost (Figure 1.3).

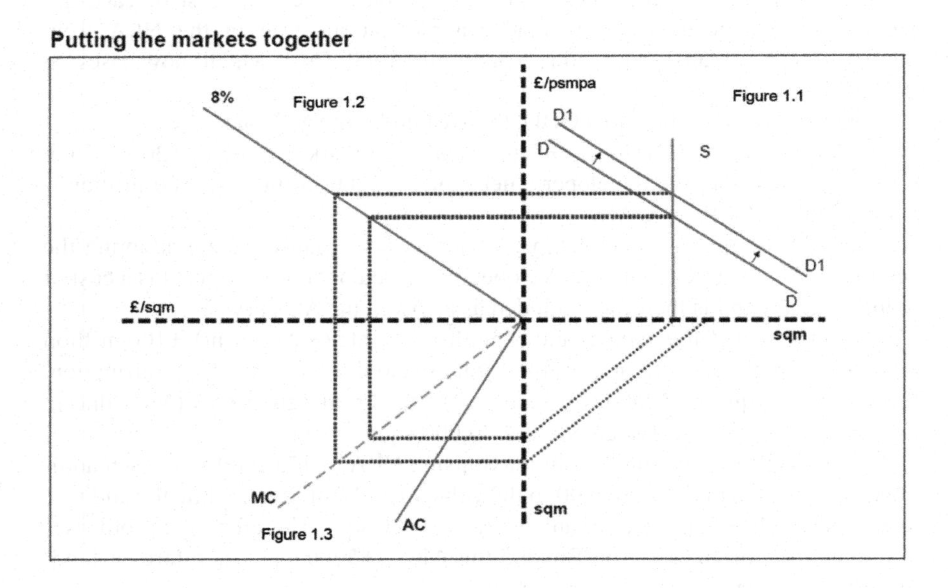

Figure 1.4 Putting the Markets Together

- **"Yield accretive" investment**: Suppose demand for space (Figure 1.1) increases, shown by a shift from DD to D1D1. Given the fixed space at 173 Pitt Street, rent will rise. Assume (Figure 1.2) that capital market conditions are unchanged, and a capitalisation rate is 8.0 percent. Then, assuming the higher rent is believed to be permanent, capital value per sqm. will also increase. Suppose the rent rises from £400 psmpa to £500 psmpa. The capital value of 173 Pitt Street per square meter is now

  £500.00 / 0.08 = £6,250 psm

Note three important qualifications to this analysis:

- This discussion has been couched in terms of NLA or physical space. It could equally be interpreted as an investment in upgrading the *quality* of the building rather than an increase in NLA. This is a likely response to a permanent rent increase if local regulations or technology limit the NLA of 173 Pitt Street to the existing 40,000 sqm. Regulations limiting the *size* or *height* of buildings are likely to result in higher *quality* buildings, or buildings with superior services – physical security or social amenities, for example.
- Note also that the analysis of the demand increase has been confined to 173 Pitt Street. If the shift in the DD line reflects a market-wide increase in the demand for space then, while 173 Pitt Street is now "under-capitalised" and ripe for expansion or refurbishment, *so too is every other office building in the City of London.* The scene is set for a market-wide cycle of demolition, construction and refurbishment.
- The assumed increase in rent from £400 psmpa to £500.00 psmpa was based on a shift from DD to D1D1 in the demand curve *with the NLA of 173 Pitt Street unchanged at 40,000 sqm.* A subsequent increase to 50,000 sqm will result in a decline in the equilibrium rent, perhaps even to below the original £400.00 psmpa. The three-market model is a *comparative static analysis* – it offers insights into one-off shifts in individual markets but does not address dynamic adjustment paths over time.

- **Buying "above market"**: For the existing owner of 173 Pitt Street, additional investment in the asset has been shown to be "yield accretive." But an alternative, and equally attractive, strategy is available. Even though capital market conditions have not changed, and the capitalisation rate of this building is 8.0 percent, a purchaser of the newly expanded office building would pay £312.5 million and earn a yield of 8 percent. However, a purchaser of the existing building would pay £256.25 million (area OFGD plus BGH). The additional area (triangle BGH)(= £6.25 million) represents the option of expanding the building by 10,000 sqm to 50,000 sqm. This payment would generate a yield of 7.8 percent on the current building, so the prospective new owner is buying "above market."

What is the relevance of this? Following the rent increase, it appears that expanding existing assets is 'cheaper' than buying at current market prices. Buildings are apparently "worth more than replacement cost." Refurbishment by existing owners looks to be "cheaper" or more rewarding than the purchase of assets because a potential purchaser of 173 Pitt Street, would have paid "over the market" at a yield of 7.8 percent. The new owner, however, would be able to assure investors and financiers that this price was financially justified. And, indeed, an expenditure of £56.25 million to increase the NLA to 50,000 would lift the yield to 8.0 percent. The "accretive" strategies of existing owners and the rationale for paying "above market" by potential new owners are both at least supported by financial analysis. They arise from the same (and external) market forces. The additional investment or transaction activity is a win-win for owners and investors, but also for market intermediaries – bankers and real estate agents.

- **The "anomaly" of simultaneously falling yields and falling rents**: An anti-clockwise rotation of the capitalisation line in Figure 1.2 to, say, 7.0 percent implies that 173 Pitt Street is "under-capitalized." This change in capital market conditions may occur independently of any shift in the demand or the supply of space (Figure 1.1). The yield applied to this property may fall for a variety of causes – a positive re-rating of the asset (or asset class) by investors or an increase in their risk appetite; a decline in the underlying real risk-free bond rate for macro-economic reasons might also support a fall in the yield. Though the initial driver is quite different from the two examples above, the impact on Quadrant 3 is similar. 173 Pitt Street is now "under-capitalised" and investment either as an increase in NLA or an improvement in building quality and amenities can be financially justified. Indeed, in a competitive market for property assets, additional investment in the building is inevitable although obviously time lags in construction must be considered.

  Demand for space, however, we assume has not changed; demand is still shown by DD (Figure 1.4). Therefore, the increased NLA in 173 Pitt Street *and potentially all other office buildings* in the City of London lead to a decline in rent. We have an apparent anomaly of a decline in yield, increased investment and falling rents. And this response is financially rational, indeed optimal.

  With the capitalisation rate falling to 7.0 percent in Figure 1.2 and a decline in rent to, let us assume, £375 psmpa, the capital value per square meter of 173 Pitt Street is now

  £375 psmpa / 0.07 = £5,357.14 psm

  As the MC line in Figure 1.5 shows, this would support an NLA increase of 4642.50 sqm.

After the NLA increase the value of 173 Pitt Street is now

£5357.14 × 44,642.50 = £239,156,125.50

or an increase of £39.156 million. Offsetting this the owner has invested an additional

£357.14 psm × 4642.5 sqm × 0.5 = £829,011.22

in the increased NLA of the building.

Figure 1.5 illustrates Figure 1.3 in more detail.

How does this increase in rent play out? A rise in rent, with a capitalisation rate of 8.0 percent implies a higher capital value. At the higher capital value (OF in Figure 1.5 = £6,250 psm) 173 Pitt Street is obviously "too small." At this higher capital value, the optimal NLA for 173 Pitt Street is now 50,000 sqm.

A 50,000 sqm office building with rent of £500.00 psmpa and capitalisation rate of 8.0 percent would have a capital value of £312.5 million

50,000 sqm × £6,250 psm = OFHK

implying an addition of £112.5 million to the capital value for the additional construction expenditure of £56.25 million (area DBHK).

173 Pitt Street is now "under-capitalised."

A rise in capital value implies an increase in supply

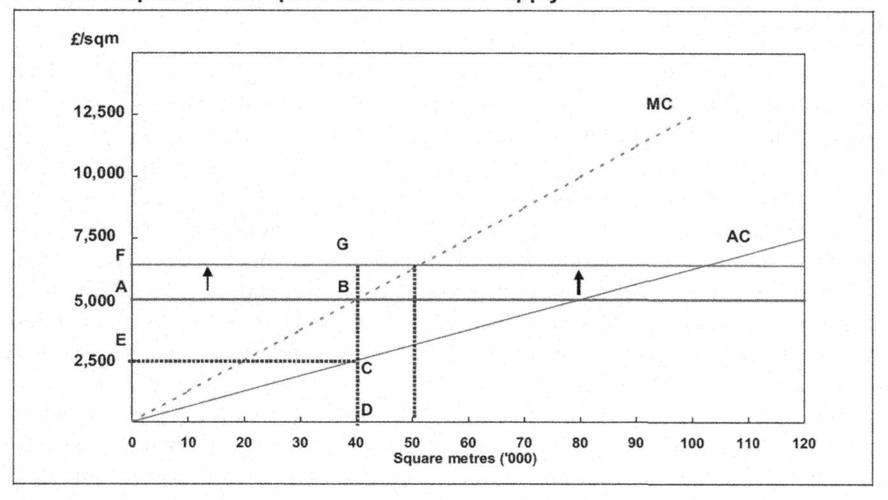

Figure 1.5 A Rise in Capital Value Implies an Increase in Supply

Obviously, this represents an attractive investment proposition. The alternative, to leave the property NLA unaltered at 40,000 sqm, would limit its value to

£6,250 × 40,000 sqm = £250 million = OFGD

an attractive but sub-optimal increase in value of £50 million.

Recall that capital market conditions, as defined in Figure 1.2, have not changed. The prevailing yield is still 8.0 percent. The expansion of the property recommends itself to the institutional owner because it is "yield accretive." Additional spending of £56.25 million (DBHK) generates additional rent of

10,000 sqm × £500.00 p.a. = £5,000,000 psmpa,

equal to area DGHK, or an "accretive" yield of

£5.0 million / £56.25 million = 8.9%

on the additional investment.

For the whole investment (£200 million initial value plus £56.25 million in construction cost = £256.25 million) the income (50,000 sqm @ £500 psmpa) gives a yield is 9.8 percent based on historical cost.

In a competitive market place, the benefits of falling yields are shared between the building owner in the short term (as increased capital value) and in the longer term by tenants (as lower rent and an increase in NLA).

Conclusion

The three markets graphical model does not aspire to explain everything about commercial property markets. Most shocks that drive real estate markets arise from outside the markets themselves – pandemics, monetary policy shifts, dot-com busts, changes in market regulation such as zoning rules and transport investment. The three markets model offers a toolkit for examining how these shocks are transmitted in various ways across real estate markets. The model also provides a coherent platform for interpreting some of the trends that can emerge during market cycles – for example, why supply might increase as rents decline; why investors might be willing to pay "over market"; why investment in existing assets is often justified as "accretive" and why simultaneously falling rents and falling yields is not necessarily anomalous.

In recent years real estate markets and their participants have been exposed to a wide range of sometimes contrary and sometimes reinforcing external influences – fluctuating demand for space across different market sectors; falling interest rates; declining equity risk premia in response to global portfolio reweighting towards real estate assets; temporary (or perhaps permanent) demand shocks from the COVID-19 pandemic. Careful analysis of these diverse and sometimes apparently contradictory trends can offer important insights and can sometimes challenge an existing market consensus – often a precondition for successful decision-making. The three-market model may not always provide clear answers to the puzzles posed by a complicated real world. But it is always a fruitful source of challenging questions.

Notes

1 See Chapters 2 and 4.
2 Fisher, J. D., Integrating Research on Markets for Space and Capital, *Real Estate Economics,* 20 (1992), pp. 161–180.
3 DiPasquale, D., and Wheaton, W. C., The Markets for Real Estate Assets and Space: A Conceptual Framework, *Real Estate Economics,* 20 (1992), pp. 181–197.
4 Chapter 4 provides an explanation of the financial arithmetic that underpins the yield (or capitalisation rate) and discounted cash flow calculations.
5 In this version of the three-market model, the supply response to fluctuations in rent reflects long-run diminishing returns based on technological or regulatory factors. The land value reflects the quasi-monopoly element in a specific location. In a comparative static or Ricardian framework it is adjustments in land value that brings the construction cost of the asset (including the developer's required or "normal profit") into line with the discounted cash value of the future income stream. Chapter 5 presents a more detailed discussion of land values.
6 For example, Baum, A., *Real Estate Investment: A Strategic Approach,* 3rd ed. (London: Routledge, 2015) (Sec 4.4.2). Chapter 4 discusses this, and other assumptions implicit in Eq (1.1) in more detail.
7 See Chapters 4 and 10 for a discussion of financial arithmetic, capitalisation rates and yields.
8 Practical factors such as planning constraints, air rights etc. certainly play a role here. However, none of this alters the fundamental proposition. If planning constraints impose *quantitative* restrictions such as a height restriction, then the adjustment will likely appear as a *qualitative* change (increased investment in fit out quality, physical security features and other amenities.). The model therefore applies equally if we think in terms of square meter "equivalents," to take account of qualitative instead of quantitative variations.

2 Debt, equity and the cost of capital

Can debt add value and enhance investment returns?

Real estate is tangible and immovable; ownership is usually clearly defined and assets are often secured by long-term contractual income streams. In liquid real estate markets, readily available transaction evidence means that the underlying value of assets can be assessed and changes in value monitored with reasonable accuracy. For all these reasons real estate assets are attractive securities to banks and other debt providers.

Debt, being typically cheaper than equity, enhances returns to real estate equity investors. More contentiously it is sometimes claimed that "cheap" debt added to the capital stack facilitates investment by reducing the weighted average cost of capital (WACC) of the enterprise, allowing equity investors to achieve desired or pre-specified hurdle rates of return on their investment.

While the arithmetic appears convincing, and there may be a good reason to include debt in a capital structure, the impact of debt on the overall cost of capital and its contribution to investment performance requires analysis. The search for the "optimal" or "appropriate" ratio of debt to equity for real estate investments, and the impact of debt finance on investment returns, are high on the financial market research agenda.

Introduction: Debt and the law of one price – the essential insight

Ms Brown and Mr Smith are next-door neighbours. Their houses are identical. Ms Brown owns her house debt-free but Mr Smith has a mortgage of £200,000.00 secured against his house. Ms Brown offers her house for public sale and after a competitive marketing process, she accepts an offer of £800,000.00.

What is Mr Smith's house currently worth?

Since the houses are identical and adjacent to each other it can fairly be judged that the market value of Mr Smith's house is also £800,000.00. This is despite the fact that Mr Smith's house is encumbered with a mortgage, while Ms Brown is mortgage-free. Mr Smith's equity in his house is, therefore, £800,000.00 − £200,000.00 = £600,000.00

DOI: 10.1201/9781003111931-2

This is an application of *The Law of One Price* which states that:

> The price of identical goods at the same point in time will be the same in a world of liquid markets, perfect and costless information and an absence of other frictions such as agency and transactions costs.

In the case of these two houses, a potential purchaser of either house will look through the financial arrangements of the current owners. The private financial arrangements of Ms Brown and Mr Smith are a matter of indifference to a potential purchaser. The successful purchaser will take responsibility for organising his or her own financing. If Mr Smith sells his house, it is he who will be responsible for settling his mortgage obligations.

Of course, multiple conditions are required for the *Law of One Price* to apply – conditions that are often violated in the "real world." Real estate markets, particularly non-residential markets, often fail to meet the perfect and costless information requirement. This is attested by the existence and activities of many commercial and industry-funded data providers, commentators and market analysts as well as resources allocated by professional investors to data gathering and information analysis. In real estate markets, information is often scarce; timely, accurate data can command a high price.

Nevertheless, the *Law of One Price* provides a useful starting point to a discussion of capital structure. And for those who are sceptical of the efficient market perspective, the *Law of One Price* is the basis of the "comparable sales" conceptual framework for deriving *market value* widely adopted as a guiding benchmark by real estate valuation professionals and market analysts.

Notice that even if Mr Smith increases his mortgage by £50,000.00 to £250,000.00 the market value of his house is unaffected. Further, if a friendly bank manager offers Mr Smith a particularly favourable deal on monthly mortgage payments this will not increase the market value of his house either – though of course, a reduction in the mortgage interest rate will be of direct benefit to Mr Smith personally. His own residual income stream will rise although his equity claim on the house remains unaffected.

Notice also that if market interest rates fall, so that residential mortgage rates are lower, the price of all houses (including Ms Brown's mortgage-free house) may rise, although Mr Smith's personal equity in his house will rise by a greater percentage than Ms Brown's. But it remains the case that both houses have identical market value. Adding debt, even "cheap" debt, to the capital structure does not increase the value of Mr Smith's house relative to Ms Brown's, which is determined by demand and supply conditions in the local residential market. Market value is determined by the demand for the asset which itself depends on the output and flow of services derived from that asset. Value in a market environment is not a "cost-plus" calculation, though there may be circumstances, particularly in the case of unique assets or illiquid markets, where this approach offers useful insights.

How does this apply in the real world?

Analysis of the impact of debt on asset, enterprise or corporate value, and the search for an "optimal' capital structure, have long been high on the financial research agenda. Few real estate transactions do not involve some analysis of alternative financing options, including access to, and cost, of debt, which is typically available in many different forms.

Every transaction is unique but it's important to get the foundation principles right. A good starting point is the path-breaking work of two recipients of the Nobel Memorial Prize in Economics Sciences, Franco Modigliani and Merton Miller (MM).[1]

The MM analysis rests on *three fundamental propositions,* all of them relevant to real estate investors:

> **MM Proposition I:** *The market value of any firm is independent of its capital structure and is given by capitalizing its expected return at the rate,* rV, *appropriate to its risk class.*

Proposition I states that in a competitive market and in the absence of extraneous (but very relevant real world) factors such as taxes, risk of bankruptcy and transaction costs, enterprises[2] that deliver identical services will sell at the same price (or rate of return) regardless of financial leverage (the debt-to-equity ratio, also sometimes described as "gearing"). *The market value of the enterprise is independent of the underlying financial structure.* The analogy with the impact of debt on the market value of Mr Smith's house is clear.

In support of Proposition I, MM offer an arbitrage pricing argument designed to demonstrate that adding debt to the capital stack, under their strong assumptions, is a zero-sum game. The arbitrage analysis is compelling, but it rests, as they indicate, ultimately on the intuitive tautology of the *Law of One Price.*[3] Proposition I has been described as "the single most important result in the theory of corporate finance obtained in the last 30 years."[4] Nevertheless, it rests on nothing more profound than the parable of the two adjacent houses.

Therefore, Proposition I, like the *Law of One Price*, is neither a *prediction* nor a *forecast* – it is a powerful *tautology* given a set of specified assumptions. It follows that attempts to "challenge" the MM position[5] are unsuccessful. Of course, it is easy to find real-world circumstances where the required assumptions set out by MM, resting as the analysis does on the *Law of One Price,* do not apply; in which case, the proposition that the homes of Ms Brown and Mr Smith have identical market values may indeed be incorrect in a specific market setting.

In fact, the assumptions, explicit and implicit, that underpin Proposition I (and the MM theory generally) are extensive.[6] A non-exhaustive list of assumptions includes:

- Capital markets are frictionless
- Individuals can borrow and lend at the risk-free rate

- There are no costs to bankruptcy
- Corporate taxes are the only forms of government levy (e.g. no personal income taxes)
- All cash flows are perpetuities (with no real growth)
- Corporate insiders and outsiders have the same information
- Managers always maximise shareholders' wealth (i.e. no "agency" costs are incurred in supervising the activities of managers themselves).

Nevertheless, the *Law of One Price* and the parable of the two houses remains a fundamental insight:

Adding debt (however "cheap") to replace equity in the capital structure of an enterprise is not an arithmetic, zero-sum transaction, though it will alter the cash flows thrown off between claimants (equity and debt providers). The impact of additional debt on financial risk and therefore the cost of equity capital must also be considered.

The value of an enterprise is determined by the earnings stream (before interest payments but after depreciation) – also described as the *Free Cash Flow (FCF)* – discounted at the rate appropriate to its risk class or industry sector. The FCF is the stream of funds available for distribution to all claimants on the company or enterprise – debt providers as well as equity investors.

Some financial definitions

Define

R	Revenue
- VC	Variable cost of operations
- FC	Fixed costs associated with operations
- dep	Non-cash accounting charges – depreciation, deferred taxes etc.

Equals

NOI	Net operating income (or Earnings before interest and taxes, EBIT)
- $r_D D$	Interest on debt (interest rate r_D, principal D)

Equals

EBT	Earnings before taxes
- T	Taxes ($T = t \times$ (EBT) where t is the marginal corporate tax rate)

Equals

NI	Net income

NI represents the accounting profit of the enterprise but it does not meet the requirements for valuation purposes. First, depreciation is a non-cash item which, while deductible for taxation purposes and relevant for assessing the performance of the enterprise, does not detract from the annual cash flow available to the debt and equity claimants. Therefore, depreciation must be *added back* to NOI. Second, if we are envisaging a stabilised asset (as implicitly we are since we are dealing here with a cash flow in perpetuity) we need to take account of the investment (*inv*) which is a negative cash flow item that exactly offsets the impact of depreciation. Therefore, the FCF thrown off by the enterprise is defined as

$$\text{FCF} = (R - VC - FC - dep)(1 - t) + dep - inv$$

Note that we add back depreciation (*dep*) which is a tax-deductible but non-cash item, and we subtract investment (*inv*) which is a negative cash flow item. In a steady-state, zero growth, enterprise (of which a stabilised real estate asset or portfolio of assets is a natural example), we assume depreciation (*dep*) = investment (*inv*).

When *dep* and *inv* net out, the FCF thrown off by the company is

$$\text{FCF} = (R - VC - FC - dep)(1 - t)$$

$$= \text{NOI}(1 - t)$$

FCF has been described as 'what's left over." It is the flow of funds available to the claimants (debt and equity providers) after all other claims for operating the enterprise (*VC* and *FC*) have been settled.

The impact of debt on the cost of equity

Consider what happens if the market value of the Brown/Smith houses rises by £100,000 to £900,000? Ms Brown's equity stake rises in value by 12.5 percent. Mr Smith's equity stake rises in value by 33.0 percent. The same arithmetic applies in the case of a decline in value of £100,000.00. Ms Brown's equity stake declines by 12.5 percent; Mr Smith's by 14.3 percent.

The volatility (and therefore financial risk) of financial returns to equity stakeholders rises with increasing leverage.

Table 2.1 illustrates an enterprise that delivers a 10 percent return on assets invested (NOI/V). The value of the enterprise (*V*) is defined as:

$$V = D + E \tag{2.1}$$

Table 2.1 The impact of a rising debt-to-equity ratio (assuming tax rate $t = 0$)

Equity (E)	*Debt (D)*	*Asset Value (V = E+D)*	*Leverage (D/V)*	*Net Operating Income (NOI)*	*Less Interest*	*Equals Net Income*	*Return on Debt* (r_D)	*Return on Equity* (r_E)
£	£	£	%	£	£	£	%	%
100	0	100	0	10	0.0	10.0	4.0	10.0
90	10	100	10	10	0.4	9.6	4.0	10.7
80	20	100	20	10	0.8	9.2	4.0	11.5
70	30	100	30	10	1.2	8.8	4.0	12.6
60	40	100	40	10	1.6	8.4	4.0	14.0
50	50	100	50	10	2.0	8.0	4.0	16.0
40	60	100	60	10	2.4	7.6	4.0	19.0
30	70	100	70	10	2.8	7.2	4.0	24.0
20	80	100	80	10	3.2	6.8	4.0	34.0
10	90	100	90	10	3.6	6.4	4.0	64.0
0	100	100	100	10	4.0	6.0	4.0	-

where:

D represents debt secured against the enterprise; and

E represents the market value of equity invested.

As with the two-house example, the value of the enterprise, V, is constant, regardless of the level of leverage, defined as D/V.

As Table 2.1 shows, the cash flow generated by the asset (NOI) is independent of the capital structure, assuming the tax rate $t = 0$. The rate of return ($r_V = 10\%$) is the underlying cost of capital. The cost of capital is the "risk-free" interest rate (often defined by the long-term sovereign bond rate) plus a premium reflecting the risk *not* of the enterprise itself but the *risk class* within which the enterprise is operating.

For example, the *risk class* may be defined as coal mining, tourism or the City of London office market.[7] Therefore, $r_V = \text{NOI}/V$ is a constant (assumed to be 10 percent in Table 2.1) regardless of the proportion of debt in the capital structure (and recalling our assumption that $t = 0$). We have also assumed that r_D is a constant 4 percent regardless of the leverage ratio. This is a strong assumption. It assumes that the cost of debt is independent of the leverage of the enterprise. Debt providers may take a different view if they believe that a higher level of debt increases the risk of default by the equity owner.

However, the return on equity ($\text{NI}/E = r_E$) rises and the rate of increase accelerates as leverage rises (Figure 2.1). This rising rate of return on equity is the additional reward that the equity investor demands (or should demand) for the increase in earnings volatility (or financial risk) associated with rising leverage. Financial risk rises because greater volatility in equity earnings follows inevitably from rising levels of debt, the servicing of which represents a fixed charge on NOI.

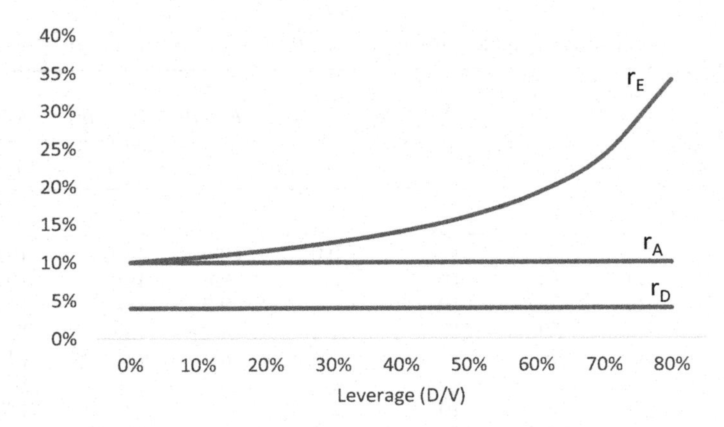

Figure 2.1 Returns to Debt and Equity as Leverage (D/V) Rises

Therefore, replacing equity with lower cost debt *does* enhance the rate of return on equity: but this is simply compensation for additional financial volatility (or financial risk). The value of the enterprise is *not* enhanced by increased leverage; debt may be less expensive than equity but it offers no 'free lunch.' Nor does adding debt to the capital structure change the investment hurdle rate or required return, r_V. The cost saving from 'cheap' debt finance is exactly offset by the additional return required by equity investors.

No matter how you choose to slice the pizza, it remains mozzarella and pineapple, with no trimmings.

This is the essential MM insight.

Why does the cost of capital or hurdle rate r_V reflect only the risk class or sector within which the enterprise operates and not the risk specific to the enterprise (say, a 40-year-old City of London office tower)?

The reason is that (in theory) an investor can choose to dilute the asset-specific risk by holding a large portfolio of assets in the office sector, diversifying the portfolio away from that specific asset or the specific location of the asset. In a large portfolio of coal mines, for example, the collapse of one mine shaft does not significantly impact overall portfolio performance. The risk of a mine shaft collapse or a fire in an office building can be mitigated by holding a large portfolio of investments in coal mines or office properties. Therefore the market does not reward the investor for asset-specific (or idiosyncratic or diversifiable or non-systemic) risk. However, investors in coal mines and office properties are rewarded for risks associated with the overall sector (such as an economic recession, unanticipated environmental legislation that limits the burning of coal or zoning changes that lead to increased construction and therefore competition for office landlords).

Similarly, for a real estate portfolio, the market value of individual assets will be discounted at the rate that reflects the non-diversifiable (or systemic) risk associated with the asset. Diversifiable risks (such as the quality of the tenants of the property, the lease expiry profile or exposure to flooding or fire risk) will diminish the market value through the impact on forecast risk-adjusted cash flows of that asset. But the discount rate, r_V, determined by the risk class of the asset, is unaffected.

The foundation MM principle, therefore, is that the cost of capital is independent of the proportion of debt in the capital structure of a company, portfolio or asset. The value of the enterprise is determined by the flow of goods and services derived from the asset, discounted at a rate that reflects the systemic (or non-diversifiable) risk associated with the business risk to which the enterprise or asset is exposed.

In practice, of course, real estate markets are cyclical. So it's helpful to consider the MM model as a long-term benchmark, making cautious adjustments to r_V to reflect prevailing market conditions.

The weighted average cost of capital (WACC)

Equation 2.2 is derived from Table 2.1:

$$r_V = r_E E / V + r_D D / V \qquad (2.2)$$

where:

r_V is the overall required return;
r_E is the return to equity investors, and
r_D is the cost of debt.

In a world where the marginal and average tax rate, $t = 0$, the cost of capital, r_V, is the weighted average of the cost of debt (r_D) and the cost of equity (r_E). That's why r_V is often referred to as the *Weighted Average Cost of Capital* (WACC).

Equation 2.2 can also be expressed as:

$$\frac{V = r_E E + r_D D}{r_V} \qquad (2.3)$$

Therefore, the value of the enterprise is the cash flow accruing to debt providers plus the cash flow available to equity investors, discounted at the WACC.

Note that Equation 2.3 is simply the conventional capitalisation rate formula found in property valuation[8] extended to explicitly identify debt and equity in the financing mix.

Since, by Proposition I, r_V is constant as leverage rises, it follows that, as shown in Figure 2.1:

$$r_{EL} = r_V + \left(D / E\right)\left(r_V - r_D\right) \qquad (2.4)$$

where

r_{EL} represents the required return to equity owners in a leveraged asset.

Equation 2.4 is also expressed as:

MM Proposition II: *The expected yield of an asset is equal to the appropriate capitalisation rate r_V for a pure equity stream plus a premium related to the financial risk equal to the debt-to-equity ratio times the spread between r_V and r_D.*

We have confined this discussion to two distinct categories of capital – debt and equity. In practice, investors and lenders contribute capital in a variety of financial instruments along a continuous risk spectrum from options, through common equity (which itself is a form of option) and progressing through preference shares to a range of debt-related instruments such as mezzanine finance to senior bank debt. The categories of capital are ranked according to their claims on the income generated and claims on the assets of the enterprise in the event of a default, with senior debt providers as the first claimant (Figure 2.2).

Mezzanine capital – typically the least secure and therefore the most expensive form of debt – is often used for short-term financing near the conclusion of a development project. Cash from pre-sales to prospective owners or investors (for example in the case of an apartment or townhouse development) during the development phase is sometimes available to developers. Although the "off-the-plan" purchasers may not see themselves as financiers, believing that they have purchased an equity stake in the completed project, they may, in fact, be participating in a form of mezzanine finance and should demand a similar rate of return.

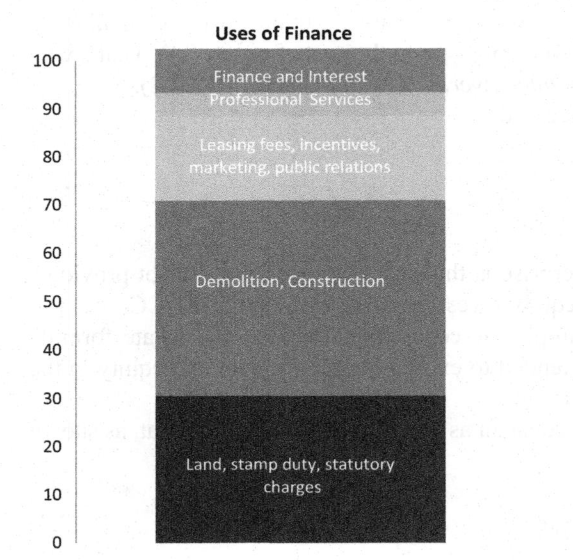

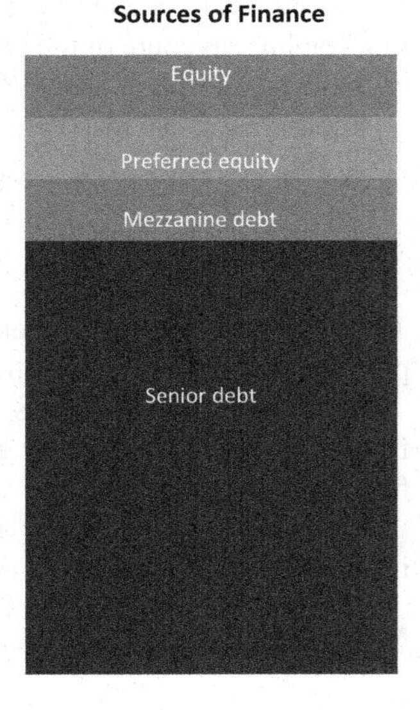

Figure 2.2 The Capital Stack – Sources and Uses of Finance

Regardless of the capital structure, however:

> **MM Proposition III:** *The cut-off point for investment for a firm (or asset) will in all cases be r_V and will be completely unaffected by the type of security used to finance the investment.*

In other words, adding "cheap" debt to the financing mix cannot be used to support a reduction in the hurdle rate applied to a prospective investment project.

MM offer the following example:

Suppose the WACC for a company (r_V) is 10 percent and the cost of debt (r_D) is 4 percent. If net income is £1000 financed entirely with equity and if income is received annually in arrears, Proposition I defines the market value V = £1000/0.10 = £10,000.

Now consider an investment opportunity requiring a £100 outlay with a yield of 8 percent. Debt financing this investment at 4 percent is apparently an attractive investment; the income of the company rises to £1,008 and the market value of the company ($D + E$) is now £10,080. However, subtracting the debt component, £100, the value of equity now falls to £9,980 as a result of this apparently profitable investment. The yield gap between 8 percent and 4 percent looks attractive, but if the market values the new investment opportunity at a 10 percent discount rate, equity investors are the losers.

Eq 2.4 shows that the return on levered equity, $r_{EL,}$ is

$$
\begin{aligned}
r_{EL} &= r_V + (D/E)\ (r_V - r_D) \\
&= .10 + (100.0/9{,}980.0)\ (0.10 - 0.04) \\
&= 10.06\%
\end{aligned}
$$

Implicit in this MM illustration is the assumption that the new investment opportunity on a yield of 8 percent is "over-priced." The appropriate risk-adjusted yield is 10 percent. MM explicitly warn that their analysis is comparative static[9] – it compares equilibria at different points in time. There is no defined "path of adjustment." In situations where interest rates or r_V can be expected to change over time (as will often be the case in long-term investments in cyclical real estate markets), MM's comparative static analysis provides a compass, not a roadmap.

It follows nevertheless that raising "cheap" debt finance as a means of achieving some predetermined "hurdle rate" necessary to convince an Investment Committee to support an investment (particularly attractive when market interest rates are near historical lows) should be viewed with caution. The reduction in the WACC (or r_V) may be an illusion unless the increase in the required return on equity, r_E, and the associated decline in the market value of E is taken into account. The low cost of debt (r_D) is readily identifiable; the rise in the cost of equity (r_E) can be harder to discern.

Similarly, companies that advertise high investment hurdle rates as evidence of a "conservative" investment policy may simply be eliminating low-risk opportunities from their menu of prospective investment targets. In capital markets, consistency, transparency and timely communication command a premium. In the absence of statements by management of investment policy, strategy, and progress (including failures) investors will always apply a discount, never a premium, in response to uncertainty.

Proposition III defines the principle that determines the investment hurdle rate, WACC or r_V. This does not imply that enterprise owners are, or should be, indifferent to the level of debt, or that the choice of capital from a wide range of alternative sources of finance is irrelevant. Other important considerations often apply.

For example, debt providers may impose, as a condition of lending, a range of constraints which limit choices or introduce costs and delays to the future decisions of management. Second the future may not be known with certainty and the cost of debt fluctuates. If, for example, owners have reason to expect market conditions to improve, they may prefer a larger exposure to debt to leverage their own rewards as equity investors in the future. Management may be awarded share options as an incentive to enhance financial performance, a further incentive to leverage up an enterprise. Conversely, if interest rates are perceived to be at a cyclical low, debt may be temporarily attractive but impose future costs as debt matures and is rolled over at a higher cost.

Therefore, Proposition III carries a warning. The practice of introducing "cheap" debt to meet some pre-selected "hurdle" rate for property investment fails the Proposition III test – "cheap" debt is in principle offset by a rising cost of equity. Equivalently, the claim sometimes advanced that an investor demonstrates "conservativism" by demanding a high investment "hurdle rate" runs the risk that this benchmark, by eliminating low-risk/low-return opportunities, will narrow the choices down to a menu of higher risk alternatives – the opposite of the advertised strategy.

Two important qualifications: The cost of debt and the impact of corporate tax

Here are *two additional important qualifications* for the MM analysis.

First, it was assumed (Table 2.1 and Figure 2.1) that bankers charge a fixed interest rate, r_D, say 4.0 percent, regardless of rising leverage. But rising leverage increases financial and default risk for the lending bank as well as for equity investors. This can be expected to be reflected in the cost of debt required by lenders. In fact, the MM model (Table 2.1 and Figure 2.1) breaks down as leverage increases – after all, at 100 percent leverage, which would be the case in a bankruptcy, the bankers effectively own the entire business. As the *de facto* 100 percent equity owners of the enterprise, the banks should be demanding a 10.0 percent return on capital invested, not the 4.0 percent required by debt providers.

However, the rising cost of debt as leverage rises *does not* undermine the MM model. MM observe however that "while the average cost of *borrowed* funds will tend to increase as debt rises the average cost of funds from *all* sources will still be independent of leverage (apart from the tax effect)."[10] In other words, the Law of One Price applies regardless of the financial structure of the asset. Under the MM assumptions, the value, V, of the enterprise is unaffected by changes in the capital structure or the cost of individual components of that capital – recall the houses of Ms Brown and Mr Smith in the introduction to this chapter.

Second, interest payments on debt may be tax-deductible to the owners if the enterprise is profitable. While corporate tax inevitably reduces FCF and therefore the value, V, of the business, the tax saving arising from interest deductibility on debt is a genuine benefit to the equity investors in the business that *does* partially compensate for the negative impact of the tax. Tax deductibility of interest payments (as opposed to the ambiguous benefits of "cheap" debt) constitutes a valid reason to utilise debt within a capital structure. The tax saving due to the deductibility of interest payments on debt represents a genuine increase in FCF that will be capitalised into the value of the business.

Cash flow after tax is:

$$\begin{aligned} \text{NOI}_T &= (1-t)\left(\text{NOI} - r_D D\right) + r_D D \\ &= (1-t)\,\text{NOI} + t r_D D \end{aligned}$$

which is capitalised so that the value of the firm, after tax, V_T, is

$$V_T = \frac{(1-t)\,\text{NOI}}{r_V} + \frac{t r_D D}{r_D} \tag{2.5}$$

Equation 2.5 assumes, reasonably, that NOI is independent of the corporate tax rate, t. Further, and arguably, the risk associated with the overall company cash flow, r_V, is unaffected by the introduction of corporate tax. Further, and mathematically conveniently, investors are assumed to assign the same assessment of risk to the "bonus" or tax shield received from the additional $t r_D D$ as applies to the expected interest payments. Therefore, r_D is the appropriate discount rate to apply to the "bonus" income stream.

Calculating V_T by Eq 2.5 and subtracting D gives the after-tax value of the equity investment V_T.

$$E_T = V_T - tD$$

So the statement that there is no 'free lunch' from increasing leverage requires adjustment – if the enterprise is liable for tax on earnings but interest payments are tax deductible, the use of debt can enhance the value of the enterprise and the WACC formula is now:

$$r_V = r_{EL} E_T / V_T + (1-t) r_D D / V_T \tag{2.6}$$

or

$$r_{EL} = r_{EU} + \left(D / E\right)\left(r_{EU} - r_D\right)(1-t)$$

where:

t is the marginal tax rate paid by the enterprise
r_{EU} is the return on equity in an unleveraged company
r_{EL} is the return on equity in a leveraged company

and, from Eq 2.3,

$$\frac{V = r_E E + (1-t) r_D D}{r_V} \tag{2.7}$$

So, applying Table 2.1, if leverage = 40% and t = 30%:

$$\begin{aligned} r_{EL} &= 0.10 + \left(0.4 / 0.6\right) (0.10 - 0.04) \times 0.7 \\ &= 12.8\% \end{aligned}$$

compared with r_{EL} = 14.0% if leverage = 40% but the tax rate t = 0.

The combined impact of corporate tax and debt is to reduce the overall cost of capital, r_{VT}, and moderate the rise in the cost of equity as the *D/V* ratio rises (Figure 2.3). The cost of equity capital rises, but more slowly. Tax deductibility of interest payments implies that the taxpayer is in effect participating in the success or failure of the enterprise. For the purposes of Figure 2.3, we retain the assumption that the cost of debt, r_D, remains independent of the leverage.

It has been emphasised that the three MM propositions are a tautology. Their direct applicability in practice depends heavily on the extent to which the "real world" satisfies the underlying assumptions. But they do provide a starting point

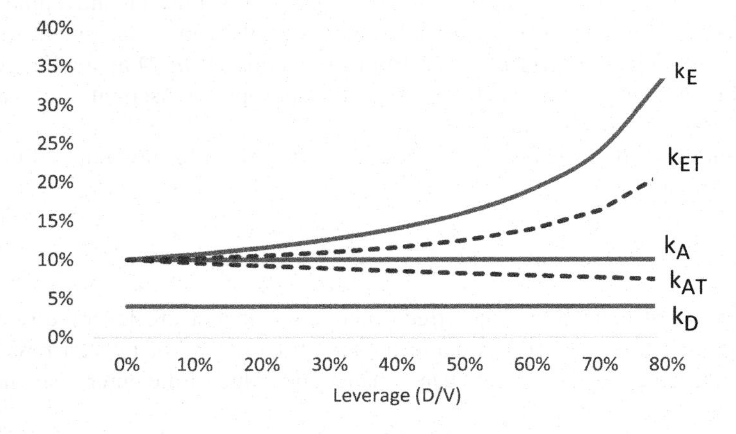

Figure 2.3 The Impact of Corporate Tax on Investment Returns

for an evaluation of alternative sources of finance, and they suggest a list of challenging questions. Like a Geiger counter in the presence of uranium, MM flashes whenever your banker introduces you to the benefits of more and "cheaper" debt and explains how an increase in leverage will enhance your investment returns, without mentioning the downside of increased volatility and the impact (if any) on the overall value of the enterprise or asset. The onus is on the banker to explain why MM's propositions do not apply, to clarify the impact of increased debt on the future volatility of returns to the equity investor and clarify the benefits to the overall value of the enterprise. As Milton Friedman once reminded us: TANSTAAFL – There Ain't No Such Thing As A Free Lunch.

Notes

1. Modigliani, F., and Miller, M., The Cost of Capital, Corporation Finance and the Theory of Investment, *American Economic Review,* 48, no. 3 (1958), pp. 261–297.
2. While the MM analysis is constructed around the value of a company, in property markets it is more natural to couch the discussion in terms of specific assets or a portfolio of assets. For convenience we shall refer to "enterprises" in this discussion.
3. Modigliani and Miller, 1958, p. 279.
4. Copeland, T. E., and Weston, J. F., *Financial Theory and Corporate Policy,* 3rd ed. (Boston, MA: Addison Wesley, 1988), p. 443.
5. For example, Baum, A., *Real Estate Investment: A Strategic Approach,* 3rd ed. (London: Routledge, 2015), Chapter 2.
6. See Copeland and Weston, 1988, p. 439.
7. See Chapter 10 for a more detailed discussion of risk class, including *diversifiable* and *non-diversifiable* risk.
8. See Chapter 4 for a detailed discussion.
9. Modigliani and Miller, 1958, Footnote 45.
10. Modigliani and Miller, 1958, p. 273, italics in original article. Similarly, a rise in the mortgage rate does not change the value of Mr Smith's house relative to Ms Jones's.

Bibliography

Baum, A., 2015. *Real Estate Investment: A Strategic Approach.* 3rd ed. London: Routledge.

Copeland, T. E., and Weston, J. F. 1988. *Financial Theory and Corporate Policy.* 3rd ed. Boston, MA: Addison Wesley.

Modigliani, F., and Miller, M., 1958. The Cost of Capital, Corporation Finance and the Theory of Investment. *American Economic Review,* 48(3), pp. 261–297.

3 The real estate Quadrant model

Finding value in the capital stack

An expanding range of investment vehicles and markets offers new opportunities and challenges to real estate investors and managers. The trend towards the globalisation of real estate investment portfolios in recent years requires that investors evaluate opportunities across locations, market sectors (retail, office, logistics, healthcare, residential…) as well as exploring the performance of alternative investment instruments (broadly categorised as debt and equity). This is an ambitious agenda. The Quadrant perspective views real estate investment as a matrix of four market categories: Public:Private::Debt:Equity.

Is real estate "different"? An international pharmaceutical, IT or automotive assembly business or, of course, a finance house may be relatively footloose. The demand drivers and the business risk are typically concentrated across global markets and international supply chain linkages; in contrast, real estate risks and returns on heavily defined by specific locations and sectoral demand drivers, requiring a specialised focus on the underlying physical assets themselves. A well-structured real estate investment process integrates traditional considerations, such as location, asset type and tenant covenants, with analysis of the relative value of alternative claims on the underlying income stream thrown off by the physical assets, alerting investors to the full range of investment alternatives, risks and rewards.

The Quadrant investment construct is not *prescriptive* nor is it a *panacea*. However, it does provide a *platform* and a coherent *perspective* to select and manage portfolios across real estate-related financial instruments: public and private, debt and equity. By focusing on the relative value between different claims on real estate assets – rather than comparing equity or debt instruments across diverse enterprises or businesses as traditionally performed by equity or credit analysts – the emphasis shifts from *relative value* across sectors and markets to an assessment of the *fundamental value* of the underlying real estate assets, their drivers and diversification opportunities. The Quadrant platform emphasises *free cash flow* and the *economic*, not the *accounting* model, of the enterprise. Emerging analysis of real-world performance suggests that investment across the Quadrants can achieve enhanced risk-adjusted return portfolio outcomes.

DOI: 10.1201/9781003111931-3

Introduction: Scanning the menu

The Modigliani-Miller (MM) theorems, described in Chapter 2, lead naturally to the contemplation of the real estate investment process. The menu of real estate investment opportunities is lengthening rapidly. In part, the expanding range of investment products is a response to the sharp increase in the volume of funds directed towards the real estate sector in recent years. The growth in market size and liquidity has facilitated new ways of packaging and securitising conventional real estate financial instruments. New real estate-related sectors, such as health-care and residential build-to-rent (also called multifamily) have emerged as institutional investment categories or, like the logistics sector, recently achieved greater prominence. A new generation of securitised products is accommodating the diverse objectives of income yield, risk mitigation and capital growth by both equity and bond fund managers. At the same time, the global economic shocks of 2007/2008 (the GFC) and 2019/2020 (the COVID-19 pandemic) are a reminder that portfolio diversification across markets and sectors has limitations as a risk management strategy; financial engineering is no substitute for traditional guidelines – cash management, debt and lease expiry profiles, portfolio diversification across markets and even within individual asset classes such as offices or retail malls. Indeed, financially engineered products can introduce a whole new set of traps for the unwary.

A further stimulus to financial innovation has been the long-term re-rating cycles of commercial property assets relative to sovereign bond markets and, recently, the compression of yields and yield spreads across the risk and duration spectrums. This long-term trend accelerated between 2007 and 2020 and, arguably, represents a new "normal." Lower yields, tighter spreads to real bond rates and a flat yield curve are all imperatives to innovation in financial packaging; they require careful analysis of headline market metrics and the fine-tuning of portfolio management strategies.

As real estate markets globalise, the range of investment opportunities expands. But the process is not without costs: greater information and research demands are now placed on analysts and investors. For example, investors and portfolio managers now need to consider exchange rate volatility and factor currency and interest rate hedging strategies into their decisions, as well as the diverse (and changing) business practices, laws, taxes and regulations that apply in different countries to real estate and related financial instruments. And, a growing range of products and strategies bridges the increasingly permeable frontier between vanilla or traditional debt and equity products, offering greater flexibility to financiers, owners and developers. Investors can now choose from a wide range of products that match their return targets, preferences for income versus capital growth and liability criteria.

More than ever, successful real estate investment requires a broad range of specialist skills, from top-down strategic perspectives to detailed product analysis. Debt markets, in particular, are growing rapidly in size and sophistication, opening a new range of opportunities to investors seeking enhanced yield. However, successful investment in these markets requires expertise in credit assessment and financial structuring as well as allocation of time and resources to monitoring and managing financial exposures. These commitments are not costless.

Easily overlooked, but no less important to the investment process, are hands-on property knowledge and day-to-day exposure to transactions in the securitised as well as the direct real estate markets. In the long-run, returns on securitised financial instruments are determined by the performance of underlying physical real estate assets – not the other way around. Globalisation of portfolios certainly offers diversification benefits, but investors ignore the traditional considerations of location, asset type and tenancy covenant at their peril.

Property investors therefore must negotiate a trade-off. On the one hand, the range of investment opportunities is expanding rapidly, both domestically and globally. At the same time, successful real estate investment dictates increased commitment of specialist resources to research, analysis and portfolio management.[1]

The Quadrant perspective divides the property market conceptually into four distinct categories (Table 3.1):

Public equity: High volatility, high liquidity, income and cyclical capital growth. The public equity market is the most liquid, transparent and heavily researched sector of the global commercial real estate market. In the United States, Real Estate Investment Trusts (REITs) and publicly listed real estate companies account for, roundly, USD 697 billion.[2] In addition to passive real estate investment, some listed real estate enterprises are engaged in a diverse range of diverse activities such as land banking, development and funds management. REITs are sometimes presented as a retail investor's proxy for private real estate investment. And over the long term, this may be correct. However, a REIT investment offers a very different package or attributes from a private equity investment. Liquidity, low transaction costs and (often) financial leverage contribute to higher levels of market volatility which can make REITs seem, at times, over-valued or under-valued relative to underlying asset values as defined by transactions evidence (which can be limited) or appraised value (which is usually informed by historical market evidence).

Table 3.1 The investment quadrants

	Equity	*Debt*
Private	Unlisted wholesale funds Unlisted retail funds Property securities funds Syndicates, joint ventures, clubs Direct property investment	Bank lending Non-bank lending Mortgage trusts Mezzanine debt Debt funds, syndicated loans
Public	Real estate investment trusts (REITs) Real estate listed companies	Corporate bonds Commercial mortgage-backed securities (CMBS) Mortgage REITs

Private equity: Low volatility, low liquidity, income and cyclical capital growth. Private equity investment has grown rapidly in recent years, as investors, both institutional and retail, have directed capital towards real estate syndicates and unlisted funds. Syndicates represent a relatively small commitment of capital overall, though these vehicles have attracted a strong following among investors seeking higher, tax-effective returns supported by relatively high levels of debt and, often, high-yielding secondary grade property. In contrast to syndicates, unlisted wholesale and retail funds are attractive because performance and capital values, being appraisal-based, are to some extent insulated from day-to-day sharemarket fluctuations. Sometimes these funds carry low levels of debt, although some private equity funds aim to achieve higher yields via the use of debt finance. Estimates of the magnitude of private equity in real estate vary widely, dependent on definitions of real estate and methodology adopted in the estimates. Roundly a figure of USD 836 billion is indicative of the magnitude of this real estate sector in the United States.

Public debt: Varying liquidity by market, stable income. Public debt comprises commercial mortgage backed securities (CMBS) and corporate bonds, usually issued by rated entities. These markets have grown rapidly in recent years, in part as falling sovereign bond yields have stimulated the search for alternative exposures to generate stable income from debt instruments. As sovereign and prime grade yields have fallen the yield gap to prime real estate debt yields has widened. In part, this reflects the underlying characteristics of real estate as an income asset, albeit often with low growth prospects. While wholesale investors are the major participants in public debt markets, the range of opportunities has been widening for retail investors. Tighter regulation of bank lending post-2008 has also widened the market appetite for public debt from a range of funding sources, while borrowers have been more focused on extending duration and building flexibility into debt maturity profiles in a way that traditional banking organisations often find difficult to accommodate. Roundly a figure of USD 1,376 billion is indicative of the magnitude of this US real estate sector.

Private debt: Low liquidity, stable income – Private debt comprises mortgage trusts and loans by banks, typically secured against specific real estate assets or portfolios of assets. The risk profile of private debt securities varies from senior debt on stabilised prime real estate assets to mezzanine finance often applied to development projects and a range of quasi-equity and convertible financial instruments. The scope and size of private debt markets vary widely between countries, with the United States being one of the most liquid and most active, reflecting both the maturity and size of the market. The low level of liquidity in most non-residential real estate markets often leads to debt instruments trading at a yield premium over bonds issued by other corporate entities. Even when apparently available, history shows that liquidity can disappear quickly in times of financial stress. The US private debt market (narrowly defined as senior whole loans) is estimated at USD 2,541 billion. Market concept is not prescriptive, but it does offer perspective.

The range of financial instruments offers choices across the dimensions of liquidity, diversification and control as well as access to diversified or niche portfolios. Binding the quadrants together are different claims on the cash flow thrown off by tangible property assets. Careful and continuous assessment of the direct property markets is a pre-requisite for successful investment in the financial instruments that derive from these assets. Subject to analysis of the underlying property markets, and assessment of the individual assets, investors in the commercial property sector must then choose from an array of alternative investment vehicles. Building on an assessment of the risks and rewards flowing from the tangible assets, investment choices require an assessment of relative values and the risks associated with these vehicles. Choices also involve a trade-off between income and capital gains. As markets become more liquid, yields tighter and the flow of funds increases, so pricing becomes finer and the quality and availability of data from the underlying direct market more important. Finally, while the quadrants can be broadly categorised along the risk/reward spectrum, this taxonomy implicitly applies to mature, stabilised assets. Development projects, for example, offer very different risk profiles that vary through the life of the project and the rising volume of capital committed as the project proceeds.

Comparing investment models

The Quadrant model: The Quadrant model emphasises the linkage between the full range of financial instruments and the underlying real estate assets from which these instruments derive their value. Fundamental to this analysis is the explicit consideration of the underlying real estate assets and the dynamics of the markets in which these assets are located. This is the first building block of the investment process (Figure 3.1).

Consistent with the economic model of the firm, the initial focus is on the volume, timing profile and volatility of the expected cash flow thrown off by the underlying assets. Once the value of the assets is established, attention then shifts to the range of claims – across the full equity and debt spectrum – that are supported by the underlying assets. It is at this stage that relative value considerations can indicate which investment is suitable for particular mandates and how the instruments issued by a particular fund or real estate portfolio rank on a risk/return basis to competing products in the marketplace.

Traditional model: By contrast, the traditional approach is based on the accounting model of the firm. Typically, credit and equity analysts independently analyse claims arising from the same underlying real estate assets. Credit analysts and equity analysts may use quite different models and valuation metrics for the same underlying assets. Investment recommendations for debt products are typically based on relative value considerations across the credit spectrum only. Criteria that influence credit rating, such as leverage and diversity of income, assume over-riding importance. Equity analysis tends to be similarly partial, with a strong focus on quantifying rental income, and a lesser focus on income quality. Each group of analysts tends to take the valuation of the other party as a "given." The resulting valuations are not necessarily constrained by an internally consistent

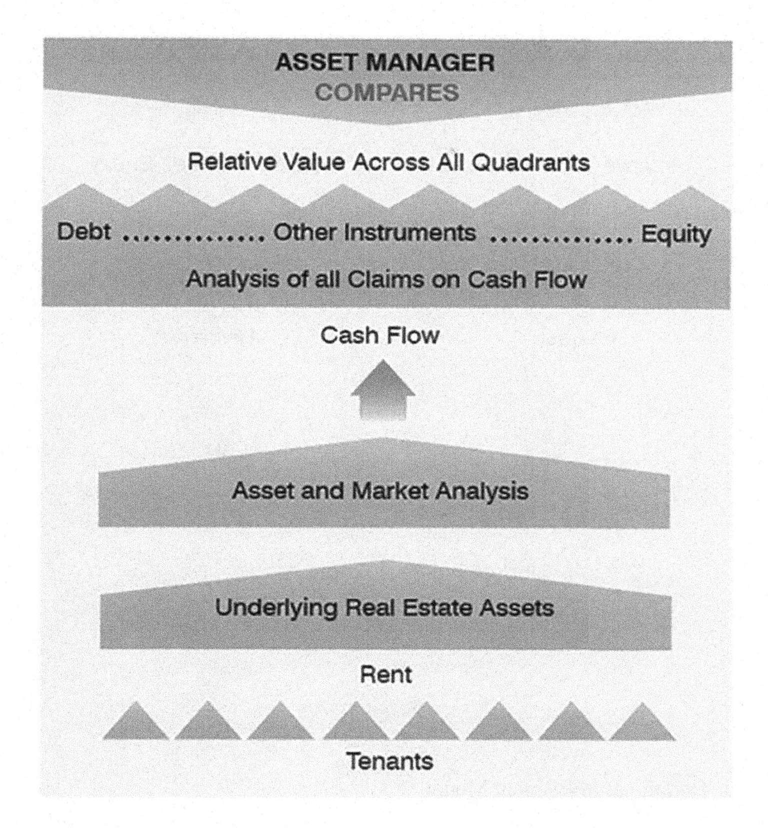

Figure 3.1 Quadrant Investment Model

model of the real estate portfolio itself, or by explicit analysis and consistent assumptions of underlying macro-economic and market trends (Figure 3.2).

In comparison with the accounting-based approach of the traditional investment model, the economic model as expressed in the quadrant approach embodies at least three advantages:

- **First,** it avoids the duplication of effort and potential inconsistency that arises when credit and equity analysts focus independently on the same underlying assets, albeit from different perspectives. Under the quadrant approach, an overall analysis of real estate assets and their underlying markets forms a necessary preliminary stage prior to analysis of the inter-related financial claims on these assets.
- **Second,** the focus on the assets, and the value of the cash flow thrown off by the assets, is consistent with the principle that investment decisions should be independent of financing decisions. A focus on the consolidated cash flow facilitates

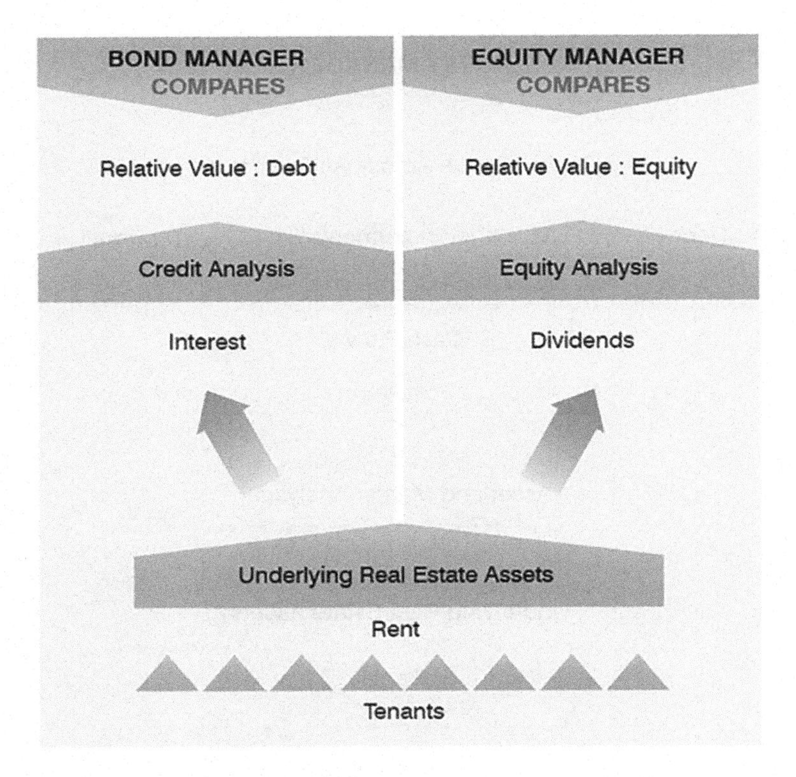

Figure 3.2 Traditional Investment Model

pricing of the underlying systemic risk associated with the assets. Accurate pricing of the cash flow at the asset level is a pre-requisite for the accurate pricing of the financial instruments that feed off that cash flow. The total value of the claims on the enterprise should equal the sum of the individual parts.

- **Third,** by looking through the traditional, and increasingly permeable, boundary that exists between credit and equity markets, a relative value comparison across all financial instruments is encouraged, without distracting and sometimes fungible boundaries between instruments defined as "debt" and "equity." Investment managers are therefore better able to match decisions with their mandates and liabilities.

The Quadrant model encourages the analyst to step away from the perspective of the conventional bond or equity fund manager. By scanning across the entire range of financial instruments, market inconsistencies and opportunities in the pricing of claims are more likely to be identified. And by considering a wider range of financial instruments the match between an investor's mandate and investment products is facilitated.

The Quadrant approach can be likened by comparing a one-stop-shop to a boutique. The Quadrant investor, in effect, is reading from the same menu as the target company's chief finance officer (CFO). Although the task of the CFO is to evaluate all sources of debt and equity capital to achieve the minimum weighted average cost of capital, the quadrant investor chooses the menu items that will deliver the highest risk-adjusted return consistent with an underlying investment mandate.

While the Quadrant approach provides a well-packaged conceptual platform for showcasing the range of alternative real estate investment opportunities, the real-world application encounters some practical difficulties. For example,[3]

- **First,** while the Quadrant approach economises on analytical resources by its exclusive focus on real estate, the task of monitoring a wide range of debt and equity markets means that its application is largely confined to large institutional investors, or boutique funds that focus on specific markets or sectors.
- **Second,** potential compliance conflicts between the information requirements of an equity investment team and the confidentiality obligations of the credit team to their client impose additional costs with the enforcement of Chinese walls, particularly as compliance rules have tightened in recent years.
- **Third,** the timing for the exit and entry points may be dictated by limited liquidity and the duration of both debt and equity instruments.

Performance analysis

Does the quadrant perspective lead, at least in principle, to enhanced portfolio performance? Evidence on this is limited by the availability of data – few real estate markets offer coverage across all four quadrants and even in these markets (the United States, Australia, and the United Kingdom, for example) historical performance data is often limited, particularly for the debt quadrants. Therefore, performance analysis is largely confined to the US market. Here the evidence, based on analysis of data from United States real estate sectors (1997–2020) is that portfolios structured across all quadrants can achieve diversification benefits and deliver better risk-adjusted returns.[4] Like all data based on historical analysis, past performance is no guarantee of future performance. And, of course, the structure of real estate markets (particularly public debt markets) has matured substantially in recent years resulting in historical data of questionable relevance to current conditions. Secondly, the data are heavily impacted by the GFC and COVID-19 shocks – which may be atypical for assessing long-term performance. The authors note the relatively strong connection between the public and private debt markets, perhaps influenced by the 2008 GFC shock.[5] Post-2008 the data suggests an increase in inter-dependence between the quadrants, reflecting perhaps the impact of a new regulatory environment, further underlining the care required in extrapolating past performance into the future.

Conclusion

The Quadrant perspective is a road map to investing in real estate. It is not a prescriptive itinerary. The Quadrant model is a way of describing the world, not changing it. The alternative investment options are identified, not mandated. The model can be interpreted as a mental construct that facilitates innovative thinking about real estate investment opportunities. By directing attention to the underlying real estate assets a measure of discipline is imposed on the investment process.

The focus on the underlying assets provides the essential building block to the valuation of the assets themselves, and forces consistency in the valuation of the financial instruments that trade-off those assets. In adopting this cash-flow-focused approach the Quadrant model aligns itself with the economic rather than the accounting corporate model.

Analysis of the full range of real estate assets facilitates the pricing of debt and equity instruments and relative price comparisons not within categories of financial instruments but between these categories. It provides a basis for *absolute* rather than *relative* value comparisons. Investors with a trading orientation may elect to take advantage of perceived valuation anomalies when they occur, regardless of whether they are in debt or equity investments, residing in private or public space. Long-term investors can more easily select from a broader range of products that matches their targets for investment return and the matching of assets and liabilities.

In effect, the Quadrant perspective challenges the existing guilds of equity and credit analysts by providing a platform for investors to trade across the full spectrum of real estate investment products. As real estate markets expand in size, sophistication and liquidity a focus on value across the full range of financial instruments becomes increasingly important. The fact that real estate debt and equity instruments sometimes display low or negative correlations, while the volatility of debt instruments is typically low in comparison with equities is an encouragement to portfolio construction within the real estate sector. The recent globalisation of real estate markets and the growth in non-core real estate sectors – such as hi-tech facilities, healthcare, self-storage, build-to-rent residential and online retailing – is a further incentive to explore the full range of risk-return options available across the real estate universe.

Notes

1 See for example Hudson-Wilson S., *Cross-Quadrant Asset Allocation*, in Hudson-Wilson, S. (ed.) *Modern Real Estate Portfolio Management* (Pennsylvania: Frank J. Associates, 2000).

2 Estimated as at June 2020. Estimates of the US market quadrant data are sourced from Pension Real Estate Association, June 2020, quoted in Farrelly, K. and Moss, A., Re-examining the Real Estate Quadrants, *The Journal of Portfolio Management,* 47(10) (2021). doi: 10.3905/jpm.2021.1.271. Outside the US and UK, few real estate markets offer the transparency or historical data to offer a comprehensive menu of investment opportunities across all four quadrants.

3 Moss, A., *Fair and Square: Why the Quadrant Model Works*, The Property Chronicle, 23 February 2021.
4 Farrelly, K., op. cit.
5 Farrelly, K., op. cit. pp. 10–11.

Bibliography

Farrelly, K., and Moss, 2021. A. Re-examining the Real Estate Quadrants. *The Journal of Portfolio Management,* 47(10) doi: 10.3905/jpm.2021.1.271

Hudson-Wilson S. 2000. *Cross-Quadrant Asset Allocation*, in Hudson-Wilson, S. (Ed.) *Modern Real Estate Portfolio Management*, Frank J. Associates, Pennsylvania.

Moss, A. 2021. *Fair and Square: Why the Quadrant Model Works*, The Property Chronicle, 23 February.

4 Inside the engine room – getting the metrics right

Discounted cash flow analysis

Discounted cash flow (DCF) analysis is at the core of the asset valuation process. The financial arithmetic of DCF analysis is a well-trodden path, but it is a path with potholes and detours. The prevalence of lease contracts (often long-term) in rented properties provides information about future cash flows; combined with low levels of liquidity in many real estate markets, this argues for the use of DCF analysis as a central source of information about market values and trends. However, real estate assets and contracts have some specific characteristics that merit special attention in DCF modelling. The fact that metrics (such as gross or face rent), definitions (such as "yields" and capitalisation rates) and contractual arrangements (such as lease agreements) differ widely between markets and jurisdictions is added reason to drill deep down into the foundations of real estate financial analysis.

This chapter introduces the standard arithmetic of the DCF model as an entry point to emphasise features particularly relevant to participants in real estate markets. These include:

- ***Yields*: why a property yield or a capitalisation rate is a "real" magnitude; therefore, it should not be benchmarked against "nominal" metrics such as sovereign bond yields or market interest rates.**
- ***Growth*: the important distinctions between *systemic*, *real* and *nominal* growth is often overlooked in market analysis and commentary; nevertheless, careful analysis of "growth" has important implications for interpreting and perhaps challenging discount rates and yield metrics.**
- ***Benchmarks*: why we (correctly) focus on *yields* as value benchmarks for real estate assets but we focus on *price-earnings (PE)* ratios, not *dividend yields*, for most analysis of share markets and corporate entities.**

All three of these topics are widely misinterpreted by financial market commentators unfamiliar with real estate markets and sometimes by real estate market analysts. Income yields (or capitalisation rates) are often used as benchmarks to assess relative value at different points in time and between property market sectors or countries. These comparisons are convenient (like share market

DOI: 10.1201/9781003111931-4

P/E ratios) but can be misleading given the many differences in lease structures and duration as well as underlying conditions such as inflation between markets and countries. The final section of this chapter introduces an algorithm for *standardising* commercial property yields through time and between markets, taking account of differences in lease duration, rent-free incentive periods, within-lease rent reviews and adjustments for nominal or real rental growth often built into lease contracts.

Introduction: Basic DCF arithmetic, traps for young players

The fundamental DCF formula is

$$PV = C_0 + \frac{C_1}{(1+r)} + \frac{C_2}{(1+r)^2} + \ldots \frac{C_n}{(1+r)^n} \tag{4.1}$$

where

PV = present value at time $t = 0$, calculated over $n + 1$ periods ($t = 0, 1, 2 \ldots n$)
r = the required rate of return, the discount rate or the opportunity cost of money[1]
$C_0, C_1 \ldots C_n$ = current ($t = 0$) and future ($t = 1 \ldots$ n) cash flows, which can be positive or negative.

Note that in Eq (4.1) there are ($C_0 \ldots C_n$) *plus r plus PV equals n+3* variables in total. If we have values for *n+2* of these variables we can always calculate the missing variable. Alternatively, if $n = \infty$ then we are modelling a *perpetuity* and we are in the realm of *Capitalisation Rate* analysis. If *r* is the unknown variable, then we are in the realm of *Internal Rate of Return (IRR)* calculations. All these metrics are derived from Eq (4.1).

In the special case where

- $C_0 = 0$ (that is, payments are in arrears, commencing at the *end* of the t = 0 or current period), and
- The cash flow is constant and of infinite duration ($C = C_1 = \ldots = C_\infty$)

then Eq (4.1) can be simplified (see Appendix 4A) to

$$PV = \frac{C}{r} \tag{4.2}$$

The simplified formula (Eq 4.2) can be extended to the case where the cash flow grows in perpetuity at a constant rate, *g*, so that

$C_1 = C_0 (1 + g)$

$C_2 = C_1 (1 + g) = C_0 (1 + g)^2$

$C_3 = C_2 (1 + g) = C_1 (1 = g)^2 = C_0 (1 = g)^3$. etc.

If growth, g, is assumed to continue in perpetuity and if payments are in arrears, that is the first ($t = 0$) payment is C_0 but paid *at the end of the first period*

$$PV = \frac{C_0}{(1+r)} + \frac{C_1}{(1+r)^2} + \ldots \frac{C_\infty}{(1+r)^\infty}$$

then, applying the method set out in Appendix 4A, Eq (4.2) can be expressed as

$$PV = \frac{C_0}{r-g} \tag{4.3}$$

Note that *payment in arrears* is unusual in real estate markets where, for example, rent is normally paid in advance. In this case, Eq (4.2) and (4.3) require adjustment. If

$$PV = C_0 + \frac{C_1}{(1+r)^1} + \ldots \frac{C_\infty}{(1+r)^\infty}$$

then by Appendix 4A, Eq (4.3) is

$$PV = \frac{C_0\,(1+r)}{r-g} \tag{4.4}$$

Evidently it is a requirement for this calculation that $r > g$. For a perpetuity to be financially meaningful the growth rate, g, must be less than the discount rate, r.

If the current or first cash payment, C_0, is known and paid at the end of the first period, and the value, PV, is sourced from, for example, public transaction evidence, the "yield," y, is calculated as

$$\frac{C_0}{PV} = r - g = y \tag{4.5}$$

if rent is paid annually in arrears and

$$\frac{C_0}{PV} = (r-g)/(1+r) = y/(1+r) \tag{4.6}$$

if rent is paid in advance.

Since in practice the industry typically derives the "yield" from Eq (4.3) without adjusting in a case where the rent payment is in advance, Eq (4.4), this means that the "yield" is often technically over-stated by $(1 + r)$.

Evidently the "yield" comprises both the required rate of return, r, and growth (if any) in perpetuity, g, although there is no direct way of separating the two variables. Transaction data limited to rent (Co) and value (PV) allows us to derive y, but the breakdown into r and g remains unknown. However, we do know that if expected growth, g rises, the observed value, y, will fall if r is unchanged.

Hidden complexities: The inflation impact

The algebra is clear. But a potential trap is concealed in this discussion.

To this point, there has been no mention of inflation and therefore no discussion of the *critical distinction* between the *real* discount rate (which we have implicitly defined as r) and *nominal* discount (which we now define as d). Since inflation has been ignored to this point, the *cash flow growth* rate, g, is implicitly a real value.

We now define the anticipated future inflation rate as f.

First, consider a world with no inflation. Let's assume $r = 8\%$ and $g = 2\%$. Since there is no inflation, g is the *real* and also the *nominal* growth rate in cash flow. Assume the first cash payment is £100, the cash flow is a perpetuity and payment is annually in arrears.

Then by Eq (4.3)

$$PV = \frac{£100}{8\% - 2\%}$$
$$= £1,666.67$$

Second, consider a world where inflation is $f = 5.0\%$ p.a., *and is expected to continue at this rate in perpetuity.*

Assuming the inflation rate f has no *real effects* on the economy (sometimes called "monetary neutrality"); therefore, the introduction of $f > 0$ leaves the "real" rate r and "real" growth, g, unchanged. However current interest rates and future cash flows are assumed to rise in line with inflation.

The wording is important here. If current and future inflation is 0 percent and at noon today, we learn with certainty (how?) that future inflation will be 5 % p.a. in perpetuity then:

- Interest rates (and bond rates) will immediately rise, because banks and other creditors will require a return on their loans and investments *adjusted for inflation.*
- Prices – of beans, electricity and the rent on office space will not change today, but in the future, they will all rise by 5 percent per annum.
- The proposition that all items – beans, electricity and office rents – will all rise at exactly the 5 percent per annum rate of inflation is open to challenge. But it is an important starting point. If you wish to make alternative assumptions about future relative prices that's OK. But you need to be explicit about your model, cite the sources and cite the supportive historical evidence.
- Since we are assuming that all items rise the same rate (5 percent per annum), it follows that real estate rents and real estate values will rise at 5 percent per annum. And since the yield (Eq (4.5)) is the ratio of rents (r) and market values (PV), both rising by 5 percent per annum, the yield will also remain unchanged.

Consider the arithmetic

If $f > 0$ the adjustment to Eq (4.3) is as follows:

The *nominal* discount rate (which we now defined as d) becomes

$$(1 + d) = (1 + r) \times (1 + f) = 1 + r + f + rf$$

Or

$$(1 + d) = (1.08) \times (1.05) = 1.134$$

So

$$d = 13.4\%$$

The (nominal) cash flow growth rate (which we now define as v) becomes

$$(1 + v) = (1 + g) \times (1 + f) = 1 + g + f + gf$$

Or

$$(1 + v) = (1.02) \times (1.05) = 1.071$$

So

$$v = 7.1\%$$

The first cash flow is now the initial payment (C = £100) paid at the *end of the first period.* The cash flow equation is now

$$PV = \frac{C}{(1+d)} + \frac{C(1+v)}{(1+d)^2} + \frac{C(1+v)^2}{(1+d)^3} \ldots \infty \tag{4.7}$$

Note that the first payment, C = £100, is received *at the end* of the first year and is therefore discounted by $(1 + d)$. The second cash flow, received at the end of the second year, is

$$C(1+v) = £100 \times 1.071 = £107.10$$

and is discounted at a rate $(1 + d)^2$.

Applying the algebra in Appendix 4A that was used to derive Eq 4.3 we derive

$$PV = \frac{C}{d-v} \tag{4.8}$$

$$= £100/(0.134 - 0.071) = £1587.30$$

The yield, $d - v = 0.134 - 0.071 = 6.3\%$.

This is the "yield" statistic that is often derived from market transaction evidence using known information on C and PV. As in the zero-inflation case, we are unable to separate the discount rate from the growth rate.

Notice that PV is slightly *lower* in the inflation case (£1666.67 compared with £1587.30) and this arises only because the capitalisation rate has risen from 6.0 percent to 6.3 percent.

Had the payment, C, been received *in advance*, however, then by Appendix 4A, Eq 4.8 simplifies to

$$PV = \frac{C\left(1+d\right)}{d-v} \tag{4.9}$$

$= £100 \times (1.134)/(0.134 - 0.071) = £1800.00$

The difference in PV between payment in advance and payment in arrears is

£1800.00 – £1587.30 = £212.70.

The difference between the PV calculations, when payment is received either in advance or in arrear, can therefore be significant. Similarly, a yield metric derived from transaction data can be significantly higher than the "correct" figure if there is no adjustment (as is often the case) for rent payments in advance.

It follows that yield statistics derived from this market evidence *without information about the timing of the initial payment* can be materially unreliable.

The penalty for payment in arrears is therefore *not only* the initial payment (£100.00) but *in addition* the 13.4 percent return on that investment over the first period which represents the time value of that payment in the first period (13.4 percent).

$£1587.35 \times 1.134 = £1800.00$

Note that the value, PV, of the asset declined from £1666.67 to £1587.30 when inflation was introduced even though the yield (8.0 percent) and real rental growth (2.0 percent) remained unchanged. This reflects the penalty that inflation imposes if the initial payment is delayed to the end of the first year *and* the rate of return on that payment over the first year.

In the case of *payment in advance*, however, the PV of the asset remains unchanged (£1800.00) when the inflation rate, f, changes, assuming that other real variables such as r and g remain unchanged.

Therefore, while Eq 4.2 (or Eq 4.5) are convenient rules-of-thumb for estimating either yield (r) or PV, they are open to significant error when inflation and/or real growth are present and when rent is paid in advance, as is often the case in real estate markets.

Pedant's corner: Why this is important?

It has been worth setting out these calculations in extensive, even tedious, detail. The reason for doing so is that it leads us to several important conclusions, sometimes ignored by market professionals, practitioners, pundits and participants.

- **Payment in advance**: In real estate markets, typically, Payments such as rent are made wholly or partly in advance. Therefore Eq (4.2), often described as the capitalisation rate formula, is an approximation. Specifically, if rent is paid in advance the first payment, £100, received by the owner at t = 0, is excluded by the formula and accordingly, the PV of the asset, if the yield, $d - v = 6.3\%$ is applied, is *understated* by £212.70 because the first payment (£100) is not accounted for and the return on that payment $(1 + d)$ through the first period is also ignored.

  But this is only the first stumbling block.

  If the market value of the asset (or a comparable asset) is known to be £1800.00 (say, from transaction evidence) then the simple application of Eq 4.3, £100/£1800.00 = 5.6% implies a *lower yield*. However, the correct yield is $d - v = 6.3\%$ (Eq 4.6). It is evident that the application of yield, or capitalisation rate as a rule-of-thumb metric, either to illustrate market trends or to compare asset values, is fraught with traps for the unwary. When yield metrics are applied to comparisons between sectors of markets where different conventions of lease duration, rent-free incentive periods, inflation adjustment conventions and economic growth profiles prevail the risks are magnified. In low-yield markets – which describes many financial and real estate markets over the period 2008 to 2020 – small differences in yield can translate into big differences in asset values. In practice, the risks attached to rule-of-thumb yield comparisons may not be trivial. If a capitalisation rate or yield derived from market evidence is incorporated in formal valuation calculations, the underlying basis for the yield calculation must be explicit.[2]
- **PV independent of inflation**: If inflation is correctly anticipated, neutral in its effects on the economy and expected to continue in perpetuity, the current asset value (NPV= £1800.00) is independent of the *future* rate of inflation *if payment is in advance*. This is because in Eqs (4.3) and (4.5) the changes to the future cash flow (numerator) are exactly offset by changes to the discount rate (denominator). It can be argued, plausibly, that the *neutral impact* of inflation on the current asset value relies on strong assumptions. However, the financial model provides a starting point for analysis; any different analysis cannot rest on simple assertion. Either an explicit statement of alternative assumptions (such as the impact of unanticipated inflation and perhaps a consequent reduction in real rental growth or an increase in the risk premium[3]) or market evidence is required. Of course, although a change in f does not, in principle, affect today's PV, nominal PV will rise faster *in the future*, reflecting the impact of inflation on rent and other cash flows.
- **Yield spread benchmarks:** The introduction of inflation at a rate of 5.0 percent per annum implied a change in the yield (or capitalisation rate) from 6.0 percent to 6.3 percent when rent is paid in arrears. Consider now what would have happened to sovereign "risk-free" long-term bond rates under the assumption of a positive inflation rate. In a zero-inflation world let us say the long-term (real

and nominal) sovereign bond yield is 3 percent. The spread between the asset yield and the bond yield is 6% − 3% = 3% or 300 basis points. This could be described as the "risk spread."[4] With future inflation of 5 percent per annum in perpetuity the long-term bond rate rises to

$$(1.03 \times 1.05) - 1 = 8.15\%$$

and the spread between the nominal bond yield and the asset yield is 6.3% − 8.15% = −1.85% or a *negative spread* of 185 basis points. If the spread between *real* asset yields and *nominal* bond rates is used as a valuation metric we must conclude, erroneously, that the asset is now substantially "overvalued." The spread between the real bond rate (unchanged at 3 percent) and the asset yield (6.3 percent or roundly 6.0 percent) however remains *relatively* unchanged. With inflation currently at historically low levels in recent years an erroneous comparison of (real) asset yields to nominal instead of real bond rates may be relatively unimportant. This may explain why the direct comparison of (nominal) bond yield or interest rates with real estate yields, and discussion about the "yield spread" between them has become so popular, even pervasive in the low inflation post-2000 world. However, analysis of historical trends shows why this analysis, based on nominal bond yields can be seriously misleading. Through the high inflation decade of the 1970s for example, nominal bond yields consistently exceeded real estate market yields by wide margins in many markets. This was not evidence of long-term asset "over-valuation" or bond "under-valuation." The "gap" simply reflected the enhanced prospect of (nominal) capital growth in real estate assets. Conversely, in a low inflation rate economy, it is tempting (but problematic) to compare (real) property yields with the nominal interest rate on bonds or bank debt. A positive yield spread may indicate an attractive cash flow in the short term but can be misleading as a guide to valuation benchmarks and to long-term investment performance, which depends on both yield and capital growth.

- **Independent" analysis:** Since Eq (4.3) is a special case of Eq (4.1) with strict assumptions, it follows that the two equations will deliver identical results *provided* that the assumptions implicit in Eq (4.3) are satisfied in Eq (4.1). This means that only in a trivial sense can the PV arrived at in Eq (4.3) be used to "confirm" or "support" or "validate" the PV obtained from Eq (4.1). If the PV's resulting from applying the two formulae are different *either* an arithmetic error has been committed *or* two different cash flows have been evaluated. The practice of using both "capitalisation" and "discounted cash flow" methods in valuations exercises, the one as a "check" or "confirmation" of the other is fundamentally flawed. The fact that in practice slightly different results are often obtained can lend spurious credence to the proposition that two "independent" methodologies have been employed. Using one presentation of the formula to verify or support the other does no more than demonstrate that minor variations in future cash flows or discount rates lead to minor variations in PV. Duplicating the calculation, albeit with minor variations, contributes nothing to validating the market value of the asset in question.

"Growth" warrants further investigation

We have considered a real growth rate in cash flows, $g = 2\%$, and in the case of inflation, $f = 5\%$, the nominal growth in cash flow is $v = 7.1\%$.

What is the source of this growth?

It may be that the growth in real cash flows arises from the underlying growth in the economy or the market sector in which the asset is located. It is, in other words, *systemic growth*. However, in this case, this growth is *already* embodied in the underlying cost of capital for the whole economy or for the sector and is therefore already reflected in the discount rate r (or d if inflation $f > 0$). Only growth in the cash flow of the specific asset *over and above systemic growth* is captured in g (or v in the case where inflation $f > 0$) and is therefore *explicitly included in the PV calculation* or in market pricing in the case of transaction evidence. But if this "growth" is reflected in the discount or capitalisation rate derived from a recent sale it is specific to that asset. It should, therefore, be applied to a different asset (e.g., an asset currently being appraised) with caution.

Any reference to "growth" must therefore distinguish between *systemic growth*, which is already reflected in the underlying cost of capital for the economy or for the asset class or market in question, and *asset or location-specific growth*. Moreover, there is reason to be cautious about claims of asset or location-specific growth in real estate markets, particularly when we are dealing with a financial model that capitalises cash flow in perpetuity. There is evidence that in at least some real estate markets *real* rental growth has been around zero over the long term.[5] Even if we ignore history, on *a priori* grounds we can expect that a long-term rise in real rents is likely to stimulate increased construction, offsetting the localised scarcity of space and limiting long-term real growth in rents. The changing skyline of the major cities of the world over the past one hundred years is an impressive testament to the strong supply response by investors and developers to expectations of higher rents arising from rising demand for office and other commercial space over the long term.

In cyclical markets, of which real estate is a prime example, rental growth is likely to rise and fall over time. In this case, we are dealing with a *disequilibrium* problem and the perpetuity DCF model (such as Eq (4.2) and Eq (4.3)) simply does not apply or should be used for valuation or benchmarking with caution. In cyclical markets, a disequilibrium model showing a market adjustment path, or explicit scenario analysis, is likely to offer important additional insights as a valuation or decision-making tool.

Sustainable growth and market metrics: Is real estate different?

Yes, it is. Investment and subsequently earnings growth in the corporate sector are often funded wholly or in part by reinvesting retained earnings. This creates a divergence between company earnings and dividend distributions. To illustrate with very broad numbers, the London FTSE 100 index was trading on a *PE ratio* of 20.3 in

December 2020 (or the inverse, an *earnings yield* of 1/20.3 = 4.92%). This implies that (again using broad numbers) the long-term required *real rate of return* (capital growth plus dividends) for the FTSE 100 sector was 4.92 percent, as of that date.

The FTSE 100 dividend yield in December 2020 was 3.42 percent, implying (broadly) that the required real return (or *capital growth*) on retained earnings is

$$4.92\% - 3.42\% = 1.50\%$$

Companies will retain dividends for reinvestment purposes so long as the anticipated real return from in-house investments equals or exceeds the hurdle rate (r) or the weighted average cost of capital (WACC).[6] Starting with the most attractive or the investment opportunity with the highest expected return, the investment will continue from one investment to the next until the return on reinvested earnings converges to the WACC. Beyond this point, the logic is to return remaining surplus funds as dividends to shareholders. Therefore 1.50 percent represents the minimum required real return on retained corporate earnings for the broad share market if the capital market is in equilibrium.

In Figure 4.1 an enterprise recorded annual earnings of £OZ per annum. Starting from O with the highest return opportunity offering a return of OB percent per annum the enterprise invests its first pound to achieve an annual return of OB percent per annum. Travelling from left to right, additional investments yield a declining rate of return to the point OW where the return on the marginal or last dollar invested is the required or hurdle return on investment, r. Investment beyond OW will lead to returns below r. Therefore, the additional earnings in excess of OW (the amount WZ) are paid out as dividends to shareholders. At the margin, therefore, or *in equilibrium*, the return is maximised when the return on retained earnings is r and surplus funds are distributed as dividends.

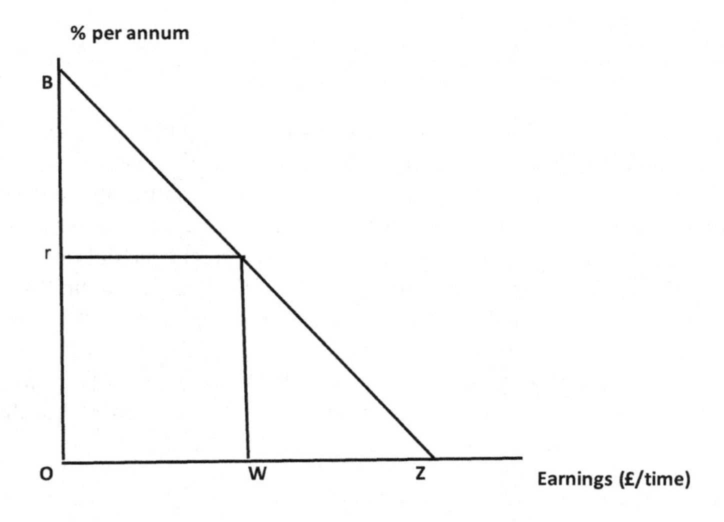

Figure 4.1 Return on Investment

From Eq (4.3)

$$r - g = C_0 / PV$$

where Co / PV is the cash yield on the investment.[7]

Therefore, at the margin (OW, O*r*)

$$r = g + C_0 / PV$$
$$= \Delta PV / PV + C_0 / PV$$

Or, required real return

r = Real Capital Growth ($\Delta PV/PV$) *plus* cash yield (C_0/PV)

But we also know that the required rate of return

r = earnings $(E)/PV = OZ/PV$

$$= OW / PV + WZ / PV$$

Therefore

$$OW / PV = OZ / PV - WZ / PV$$
$$= OZ / PV \left(1 - WZ / OZ\right)$$

or

$$g = r\left(1 - p\right)$$

where p is the payout ratio (the proportion of earnings paid out as dividends), WZ/OZ.

In equilibrium therefore growth (g) is equal to

return on investment (r) × (1 *minus* the payout ratio)

This is an important result for all financial analyses. It defines the sustainable growth rate of an enterprise. The identity applies in either *nominal* or *real* terms, if the same definition is consistently applied to r and g. The example above is defined in real terms A commercial enterprise such as a retailer or a manufacturer may retain a portion of earnings to fund future growth. Therefore, the payout ratio $p < 1.0$.

How does this apply to property investment? In the case of a stabilised real estate asset such as an established office building or shopping mall, the scope for additional real growth (in perpetuity) arising from additional physical investment that drives increased earnings does not typically exist. You are unlikely to retain a portion of this year's earnings to add another level to your office building or shopping mall. And recall that earnings are defined *net of depreciation and tax.*

Therefore, for stabilised real estate assets it is generally the case that the payout ratio, $p = 1.0$, and therefore $g = 0.0$.

In real estate metrics, therefore, we typically report the dividend or rental yield rather than the PE ratio or earnings yield, as is the case with, for example, stock market metrics.

Unlike corporate entities in other sectors, in real estate net investment leading to real earning growth in perpetuity is zero. By contrast in other market sectors reinvested retained earnings can, and often do, contribute in an important way to future earnings growth and the PE (or its inverse, the EY ratio) which incorporates both the cash yield *and* capital growth, provides a more meaningful measure of future total returns.

Cross-border valuation benchmarks – Establishing a common benchmark for comparison

Compare the pair.

- In the Sydney CBD office market in Q4 2019 the typical prime office lease incorporated a rent-free incentive period for the first 25 months of a ten-year lease. The typical prime office lease also incorporated an annual rent escalation of 4.0 percent per annum, compared with the prevailing inflation rate of 1.8 percent and a forecast long-term inflation rate of 2.35 percent.
- By contrast, in the Singapore prime office market on that date the typical lease term was three years, with no rent escalation or rent-free period. The forecast long-term inflation rate was 1.70 percent.

At this time the Sydney CBD prime office yield was assessed at 4.50 percent and the Singapore prime yield was 3.55 percent.

Are these yield comparisons meaningful?

Given the notable differences between the lease structures and economic conditions in different markets, a direct comparison of yields as a valuation benchmark looks risky. Recall that yields, derived from market transaction evidence of PV and C incorporate the hidden values of g and f, which we know vary widely between countries.

Like most summary market metrics (such as PE ratios in share markets) convenience is traded off against precision. Nevertheless, yields are one of the most widely used value metrics in real estate, even though they can mean very different things when comparing markets, market sectors or countries.

Transparency and consistency are particularly important under current market conditions when yields and yield spreads (between markets and relative to sovereign bond rates) are often close to all-time lows. Under these conditions, headline yield statistics can give a false impression of market rankings and relative value. A consistent yield metric should take explicit account of important differences between markets and even through market cycles in the same market or sector. In defining a consistent basis for yield comparisons, particular importance should be attached to differences in

- Lease duration
- Rent-free initial incentive periods (if any)
- Long-term expected rental growth rates and treatment of reversions on lease renewal and within-term rent resets
- Long-term expected inflation rates and whether these are built into lease contracts
- Adjustments for current market conditions (which may be atypical, for example after an economic shock such as the Global Financial Crisis) or some other event (e.g. the COVID-19 pandemic) and long-term "typical" conditions beyond the term of the current lease.

As an illustration, Figure 4.2 shows a yield comparison across a range of office markets as of March 2015. The Market Yields are the pro-forma quoted yields in individual markets. The Adjusted Yields standardises the yields by Eq (4.3A) to facilitate like-for-like market comparisons.[8]

This standardisation process assumes an investor acquires an existing asset, leases it on current market terms for the duration of the current lease (which varies between markets but is fixed for each individual market) and then re-leases the asset in perpetuity on market terms (which may differ from the

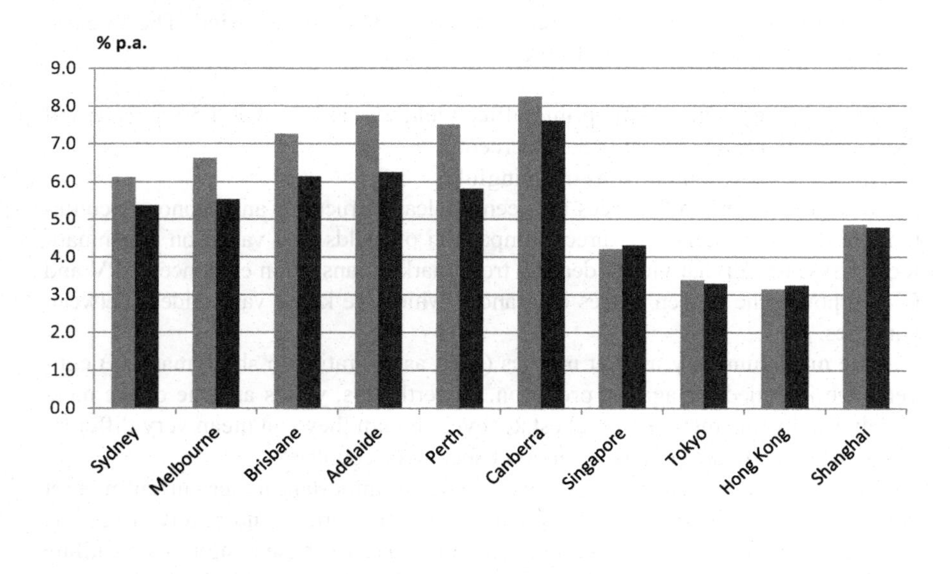

Figure 4.2 Pro Forma Market and Adjusted Prime Office Market Yields, March, 2015

current market, but are assumed to revert to "normal" market conditions for all future leases. The asset remains fully leased. The adjusted yields then measure the real rate to return to the investor.

Appendix 4A – Deriving the Capitalisation Rate Formula

First, some algebra. Define S as the sum of the infinite expression

$$S = w + wk + wk^2 + wk^3 + \ldots + \infty \tag{4.1A}$$

Then multiplying both sides by k

$$Sk = wk + wk^2 + wk^3 + \ldots + \infty$$

Subtracting the second equation from the first

$$S(1 - k) = w$$

Therefore

$$S = w/(1 - k)$$

Now apply this to property values. Referring to Eqs (4.1) and (4.2), let $PV = S$, $C = w$ (a constant term) and $k = 1/(1 + r)$

Since $(1 - k) = r/(1 + r)$ it follows that

$$PV = C(1 + r)/r \tag{4.2A}$$

However, if the first w term is removed so that

$$S = wk + wk^2 + wk^3 + \ldots + \infty$$

Then, repeating the manipulation

$$PV = C/r \tag{4.3A}$$

which is Eq (4.2) where PV is the Present Value, C is the cash flow (or rental income per period) and r is the discount rate.

Removing the first term w from Eq (4.1A) is equivalent removing the first cash flow term (C_0) in Eq (4.1), which means that the first rent payment is one period *in arrears*. Conversely, if rent payments are in advance (as is often the case in real estate contracts) with the first payment immediately, Eq (4.2A) applies.

Therefore, the summary metric (Eq (4.3A) and Eq (4.2)) are correct *if the first payment is at the end of the first period* (in arrears).

Further, if we define $k = (1 + g)/(1 + r)$ where g is *real growth* in rent and r is the *discount rate*, then

$$PV = C/\left(r - g\right)$$

which is Eq (4.3). This is the fundamental *capitalisation rate formula* employed in real estate.

Appendix 4B – Standardising cross-border yield metrics for a market or asset

Assuming we know the current Rent (C) and current Value (V), let T be the standard duration of a lease in the market or sector. Let T_0 be the term of the current lease and $T_{1\,.....\infty}$ be all future leases, in perpetuity. This allows for the possibility that current market conditions are abnormal but, by assumption "normal" conditions will reassert themselves beyond the term of the current lease. If current market conditions are "normal" then

$$T_0 = T_1 = \ldots T_\infty$$

S_0 is the present value (V) of the current lease and S_1 is the present value of all future leases in perpetuity. The distinction between S_0 and S_1 makes allowance for current, perhaps atypical, market conditions and, in the long term, a return to "typical" or "normal" market conditions beyond the expiry of the current lease.

Then extending the methodology explained in Appendix 4A, we can derive

$$PV = S_0 + S_1 \qquad (4.4A)$$

where

$$S_0 = (C\left(zb\right)^A (1 - zb(T - A)) / (1 - zb)$$

and

$$S_1 = (Cz^{(T+B)}k^B(1 - (zk)^{(T-B)}) / ((1 - z^T)(1 - zk))$$

Then, depending on standard lease conditions in individual markets,

C = Current rent
T = duration of lease (say, 120 months for a ten-year lease)
A = rent-free incentive period in current lease (say, 36 months' rent free)
B = rent free incentive period in all future leases (say, 24 months' rent free)
b = escalation in current lease (say, 4.5% p.a.)
k = escalation in all future leases (say, 4.0% p.a.)

and we define

$$z = ((1+g)\times(1+f))/(1+d) = (1+g)/(1+r)$$

where

g, f, d and r are as previously defined

Then, since C and PV are known, we solve for r (using, e.g., the MS Excel "Solver" algorithm). This is the annual average real return for an investor in perpetuity subject to the input assumptions. Since this is calculated as perpetuity the impact of the lease duration is standardised, and Eq (4.4A) adjusts for the impact of T, A, B, b and k during the current lease terms (S_0) and the impact assumed long-term "normal" lease terms in perpetuity (S_1) beyond the current lease term. Implicitly current market conditions may be "abnormal" and these drive S_0. Beyond the current lease term, the future is uncertain; therefore, we apply the long-term or "typical" lease terms to calculate S_1.

Notes

1 This is discussed in more detail below and in Chapter 2.
2 The same qualifications of course apply to the use of Price-Earnings (PE) multiples as valuation benchmarks in equity markets.
3 These matters are discussed in, for example, Friedman, M., Inflation and Unemployment, *Nobel Memorial Lecture, December 13, 1976* (Chicago: University of Chicago, 1976).
4 See Chapter 4.
5 For example, Wheaton, W. C., Baranski, M. S., and Templeton, C. A., 100 Years of Commercial Real Estate Prices in Manhattan, *Real Estate Economics,* 37, no. 1 (2009), pp. 69–83.
6 See Chapter 2.
7 Note that we could have used Eq (4.5) where $d - v = C/V$. Since d and v are both nominal magnitudes the inflation impact is eliminated in $d - v$.
8 Source: JLL Research, *Global Office Market Yields – Updating the Cross-Border B*enchmarks (s.l.: s.n, 2020).

Bibliography

Friedman, M. 1976. *Inflation and Unemployment, Nobel Memorial Lecture, December 13, 1976.* Chicago, IL: University of Chicago.

JLL Research. 2015. *Yield Comparisons for AP Office Markets - Standardising the Cross-Border Metrics.*

Wheaton, W. C., Baranski, M. S., and Templeton, C. A. 2009. 100 Years of Commercial Real Estate Prices in Manhattan. *Real Estate Economics,* 37(1), pp. 69–83.

5 Decisions, decisions … trick or treat?

Real options and the price of choice

Long-term financial commitments, high transaction costs, exposure to changing financial trends and unpredictable economic shocks…these all characterise real estate markets. Decisions to purchase, develop, refurbish or demolish assets often involve big initial outlays by owners and investors, with uncertain outcomes. A signed lease, likewise, is a mutual commitment by a tenant and a landlord, often for an extended period. Choices, once made, can be costly, even impossible, to reverse.

Discounted cash flow (DCF) analysis, discussed in Chapter 4, lies at the core of real estate decisions. For investment decisions, the DCF model advises:

***Accept* a project if the Present Value (*PV*) ≥ £0.00**
***Reject* a project if Present Value (*PV*) < £0.00**

However, DCF has limitations as a decision tool. DCF analysis presupposes a binary either/or decision at a point in time. The real world is seldom so simple. Enter, *real options.*

Real option analysis is particularly appropriate for real estate markets. In these markets information is typically scarce, transaction costs are usually high and long-term commitments are often required. At the same time, the assets themselves are immovable and therefore vulnerable to unanticipated changes arising from a range of economic, financial and regulatory sources – local and global. Therefore the flexibility to modify plans, postpone action or even reverse decisions as circumstances change can have significant financial value. It is this value that real option analysis seeks to identify.

Real option analysis of a project can uncover hidden value and suggest strategies that orthodox DCF analysis often conceals. Real option analysis can provide critical insights and suggest alternative opportunities, particularly when the future is uncertain, as it always is. Real option analysis is complementary with, but not an alternative to, DCF analysis.

DOI: 10.1201/9781003111931-5

Introduction – assessing the value of indecision

The physical and financial features of real estate are inescapable. Decisions can be costly to make, sometimes even more costly to reverse. Low levels of liquidity in real estate markets can be a challenge for portfolio managers. For example, it is estimated that in 2016 the percentage of real estate trading in the US market was around 6 percent of market capitalisation, 4 percent in the UK and Germany, just over 1 percent in China.[1] In comparison, turnover as a percentage of market capitaliation in public equity markets in 2016 was 95 percent (United States), 75 percent (Germany) and 250 percent (China).[2] Low liquidity makes the identification of market values and trends problematic and undermines the accuracy of published market data.

Transaction costs in real estate markets are often high, emphasised by a scarcity of publicly available market information and, in many jurisdictions, taxes and levies applicable to real estate transactions. Therefore, the costs and irreversible nature of many real estate investment decisions can make *optionality* that is often built into contractual arrangements, extremely valuable. Often, though, the value of an option is not transparent; the value is implicit in the negotiation strategies and financial calculations of the contracting parties.

An investment committee is often seen as a forum to share information, analyse data and recommend action; decisive decisions are one metric of the committee's success. However, a decision *not* to decide can sometimes be more valuable than a definite choice to either commit to, or to abandon, a project. Sometimes a material positive value can be attached to a decision postponed or a project modified in light of new information.

Assessing an explicit dollar value of an option (even if this is only an approximation) can facilitate the negotiation process and enhance the bargaining position of a contracting party. Options of this type are called ***real options***.

Real and financial options

A ***real option*** confers the right, but not the obligation, to undertake, amend or terminate certain business activities or investments over some defined time period in the future. ***Real options*** typically apply to real-world decisions involving physical activity and tangible assets such as real estate development projects, leases or rights to acquire or dispose of assets.

By contrast, a ***financial option*** is a financial instrument.

In Chapter 2 we illustrated how in a world of efficient financial markets, as assumed in the Modigliani–Miller analysis, for example, corporate financial decisions neither create nor destroy value. These decisions can in principle be reversed or replicated by investors with access to financial instruments. This is the province of ***financial options***.

In contrast, ***real options*** involve decisions that cannot be undone or reversed in financial markets – decisions to abandon or accelerate a construction project, for example. These decisions do create or destroy real value.[3]

Real option analysis is an important financial tool, complementary to standard DCF analysis, that seeks to assign an explicit value to flexibility. Real options can demonstrate the value of alternative management decisions at critical times through the life of a project. The orthodox PV rule, by contrast, implies a passive role for management once a buy-or-sell decision has been made.

If DCF analysis is analogous to an Exocet missile – once committed there is no turning back – real options are a drone – the pilot is in control up to the point of impact.

While optionality is frequently implicit in real estate negotiations and pricing, explicit calculation of the value of the optionality is rarely available. Nevertheless, it can be asserted that the value of a real option increases with the uncertainty and complexity of the project and the costs involved in committing to a final course of action. Real options combine flexibility, leverage and an assessment of risk to create value.

Real options can be applied in many contexts in real estate – examples include

- **Investment**: to invest or not to invest in a project
- **Urgency**: to accelerate or slow down a long-term project
- **Delay**: to invest now or defer to later
- **Flexibility**: to terminate, modify or extend a lease contract
- **Suspension**: to mothball an asset or terminate an activity indefinitely until market conditions improve
- **Abandonment**: to demolish an asset or finally terminate a project

These are real-world decisions that have financial payoffs.

Real options analysis seeks to attach dollar values to these choices. The diversity of option situations means it is difficult to provide a definitive list of real option models as they apply to real estate – but the broad principles are readily established.

What's wrong with the present value (PV) rule?

Nothing is "wrong" with the PV rule; but it has limitations, particularly when applied to projects involving high levels of uncertainty around the timing of cash flows and decisions involving up-front and future commitments.

The simple PV rule for an investment decision is:

Proceed if $PV \geq £\ 0.00$
else
Do Not Proceed.

The PV rule sets out a decision rule at a particular point in time. Beyond this, the rule implies a passive role for management. But real-world decisions are seldom as simple as this.

In a world of uncertainty and contingent management decisions, the PV rule can be wrong – sometimes very wrong. Indeed, *the decision to delay or defer a decision* can be one of the most valuable outcomes of an investment analysis process. Analysis of *optionality* can often add substantial long-term value to a project or turn an apparently loss-making project into a profitable venture.

For example, choices for a developer may be:

Stage 1: secure the site, obtain the necessary planning approvals and identify sources of finance OR pass up the opportunity

Stage 2: demolish the existing building and clear the site for construction OR delay and continue to collect the rent

Stage 3: commence construction now OR postpone the development until market conditions improve

Stage 4: review market conditions and commence construction OR once again postpone the proposed project OR abandon the project permanently and sell the site.

At each stage, there are cut-offs to the development process, and because decisions are made sequentially over time, market conditions, information and economic circumstances can change. Some decisions can be revisited (such as the decision to postpone development) whereas other decisions (such as demolition of a building) are irreversible.

The problem is to explore alternative scenarios and assign a monetary value to flexibility. This is the realm of real options analysis.

Decision deferred – the basic principles of real options

Whereas financial markets often support public markets for traded options, and the methodology for pricing financial options is largely agreed – for example, applications of the Black-Scholes option pricing formula – real options are less widely known and applied. And, real options are particularly relevant to real estate decisions.

Although applied in different contexts, real option pricing is based on the same principle as applies to financial options:

***The fundamental Option Pricing Principle**: If we can identify or construct an actual or hypothetical financial instrument that delivers outcomes identical to the proposed project or investment under different scenarios, and if we can determine the value of this financial instrument, then we know the value of the option.*

Consider a landlord's conundrum.

One strategy to attract, or to retain, a tenant might be to offer a reduced rent or a generous up-front rent-free incentive period. Both these strategies imply a direct cash cost to the landlord.

An alternative strategy might be to offer contract optionality. Future choices about lease terms, flexibility on space occupied or additional services provided by the landlord through the term of the lease may be attractive to the tenant – something the prospective tenant may be willing to pay for. These options have a value, and in markets and circumstances where uncertainty is high, the value attached to optionality by a prospective tenant can be very great. This strategy might be a cash increase to the landlord

To establish the basic arithmetic that illustrates the value of an option consider a simple case[4]:

Suppose a share in a company can be purchased now (t = 0) for £20.0 (S = £20.0).

At the end of the first period which we assume is one year hence (t = 1) the share price will be either

Up case: $S^* = £24.0$ where $S^* = uS$ and $u = 1.20$

or

Down case: $S^* = £13.4$ where $S^* = dS$ and $d = 0.67$

To keep it simple, the share pays no dividends and transaction costs such as brokerage are zero. We can borrow or lend at a rate of 10 percent per annum.

At $t = 0$ we can purchase a call option (C) on this stock with the expiry date $t = 1$, one year hence, and a strike price of $K = £22$, noting that the current share price, $S = £20$.[5]

At t = 0 we then consider the Up (Cu) and Down (Cd) scenarios for the call option:

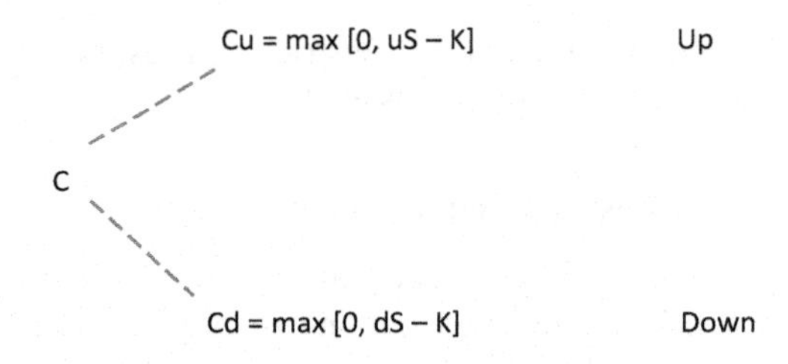

At $t = 1$ there are two alternatives:

Up case:

$S^* = uS = £24.00$ and the value of the call option with a $K = £22$ strike price is

$$Cu = £2.00,$$

or

Down case:

$S^* = dS = £13.40$ and the value of the call option with a $K = £22$ strike price is

$$Cd = £0$$

So far we do not know the price of the call option, C, at $t = 0$.

To determine C we set up a hedge strategy now ($t = 0$) so that the investor is *indifferent* at time $t = 1$ between the Up and Down case.

Suppose at $t = 0$ we buy 1 share at the current price, £20, borrow £12.18 at the interest rate of 10 percent per annum and write (sell) 5.30 call options with a £22.00 strike price.[6]

At $t = 0$: Set up the hedge. Buy one share for £20.00 (cash outflow) and borrow £12.18 (cash inflow). Sell (or "write") 5.3 call options[7] with a £22.00 strike price (cash inflow). If the net cash flow of these three transactions is zero, the value of 5.30 call options must be £7.82 (or £1.48 per option).

At $t = 1$: There are two alternative outcomes.

In the Down case, one share is worth £13.40 and the cash repayment, including interest at 10 percent per annum is £12.18 x (1.10) = £13.40. The call option is worthless (£0.00).

In the Up case, the share is worth £24.00, and the cash repayment including interest is £13.40 but now the investor must buy 5.30 calls at £24 *minus* £22 = £2.00 each to settle the short sale. This costs 5.30 x £2.00 = £10.60.

The net cash flow is identical (£0.00) in both the Up and Down cases, reflecting the perfect hedge.

In a market for financial options, a call option at $t = 0$ would trade at £1.48. *This must be the value.*

Why?

Suppose the option value were lower, say £1.20. We could then sell 6.52 call options (6.52 × £1.20 = £7.82), buy one share (−£20.00) and borrow £12.18. At $t = 0$, net cash outflow of £0.00.

Then at $t = 1$

Up case:

Share: £24.00
Repay debt: −£13.40
Option value: 6.52 × £2.00 = −£13.04
Net cash flow: −£2.44

Down case:

Share: £13.40
Repay debt: £13.40
Option value: £0.00
Net cash flow: −£0.00

Since we would lose £2.44 on this transaction in the Up cases and break-even in the Down case we would not write (sell) an option for £1.20.

Conversely, had the option price been higher (say, £1.60) we would sell 4.88 call options for 4.88 × £1.60 = £7.82. Then in t = 1
Up case:

Share: £24.00
Repay debt: −£13.40
Option value: 4.88 × £2.00 = −£9.76
Net cash flow: £0.84

In the Down case, as for the Up case, the net cash flow is £0.00. However, since we can lock in a risk-free profit in the Up case and break even in the Down case the number of options written would increase, driving down the price to £1.48.

At this point it is worth restating the **Option Pricing Principle**: *If we can identify or construct a hypothetical financial instrument that delivers outcomes identical to the proposed project or investment under different scenarios, and if we can determine the value of this financial instrument, then we know the value of the option.*

Notice, importantly, that in this analysis we have not made any explicit assumption about the probability, p, of an Up outcome and, equivalently, the probability $(1-p)$ of a Down market.

Had we, for example, assumed $p = 0.60$, then $(1-p) = 0.40$ and, applying the NPV rule:

$$\begin{aligned} PV &= -S + \left(p \times uS + \left(1 - p\right) \times dS\right)/\left(1 + r\right) \\ &= -£20 + \left(0.6 \times £24 + 0.4 \times £13.40\right)/\left(1.10\right) \\ &= -£2.03 \end{aligned} \tag{5.1}$$

The assumption, $p = 0.60$, might have been based upon analysis of historical data or the collective judgment of the investment committee. Suppose we had selected $p = 0.50$? Then

$$\begin{aligned} PV &= -£20 + \left(0.5 \times £24 + 0.5 \times £13.40\right)/\left(1.10\right) \\ &= -£3.00 \end{aligned} \tag{5.2}$$

And it remains the case that $PV < 0$.

Since $PV < 0$, the PV rule advises against this investment now ($t = 0$). However, this takes no account of the flexibility available in an option strategy.

We now conduct a mental experiment.

If you are required to make an investment decision *now*, your decision will be to *not* invest in the project, by the binary PV rule.

However, suppose you have the right (the *option*) to defer your decision by one year? How much is this option worth? The current owner of the asset (or the project) asks the prospective investor: "How much will you pay *now* if I allow you to defer your decision until this time next year?"

The benefit of this proposal is that one year from now the investor will know which of the two states, the Up or the Down cases prevails.

The value of flexibility is the difference between PV without flexibility (–£2.03) and the option value already established (£7.82).

Since the NPV of the delayed decision (£7.82) exceeds the PV of the decision to proceed immediately (–£2.03) the correct choice is to postpone a decision until $t = 1$. How much is this flexibility in the timing of the decision worth?

Additional value of flexibility = Value of option *minus* PV without flexibility

$$£7.82 - \left(-£2.03\right) = £9.85 \tag{5.3}$$

Note that instead of borrowing £12.18 from a third party the investor could have self-funded it, receiving a 10 percent return (£12.18 × 1.10 = £13.40) in $t = 1$, equivalent to the rate of interest for a riskless hedge.

Three important observations

First, although we have not made any assumption about the probability, p, that the Up case will occur, the probability value is nevertheless implicit in the values of S, U, d and r. We can also express the call option value C as

$$C = \left[p\ Cu + (1-p)\ Cd\right] / (1+r)$$

where

$$p = (1 + r - d) / (u - d) \tag{5.4}$$

Applying the values from Table 5.1, $p = 0.81$. In Eq (5.1) it can be shown that if $p = 0.81$, NPV = £0.00 and the investor is indifferent as to whether to invest now or to postpone the decision. In other words, the investor is fully hedged. The value $p = 0.81$ is therefore embedded in the option pricing formula.

Note that $p = 0.81$ is the *risk-neutral probability*. The risk-neutral probability is *derived from* the pricing of the call option, C. It is not an *input into the calculation of C*. This is because we are assuming that the existing market, with S = £20.00, already incorporates investors' risk preferences and other factors that might influence the share price. For example, suppose (as an extreme case) $p = 1.0$. Then S is certain to be £24 at $t = 1$. Therefore, the current price of S must be £24.00/1.10 = £21.82. This alters u (which is now equal to $(1 + r)$) while d is irrelevant because the Up outcome is known with certainty.

This is reflected in today's market asset prices – more risky assets sell at a discount to less risky assets. And, recall, it's *relative asset prices* that matter. But suppose we consider an investor who is "risk neutral"? This investor would assess the market price of "risky" assets as too low – a "buy" opportunity in other words.

What would be the Up and Down probabilities *given current asset prices?* Equation (5.3) answers that question.

Another way of looking at risk neutrality is as follows: Investors are assumed to be risk averse. This does not mean that they prefer to avoid volatile assets. For investors, the *risk* of an asset is defined not as the volatility of the asset, *but by the contribution of that asset to the stability of a portfolio comprising all available assets.* A focus on overall portfolio performance is the central proposition of the

Table 5.1 Option value arithmetic

Cash flow (£) Inflow(+)/Outflow(−)				
		t = 0	Payoff (t = 1)	
			Down	Up
1	Share	−20.00	13.40	24.00
	Borrow	12.18	−13.40	−13.40
5.3	Call Options	7.82	0.00	−10.60
	Net Cash Flow	0.00	0.00	0.00

Capital Asset Pricing Model.[8] Therefore "risky" assets are those investments that are defined as assets that contribute to overall portfolio volatility. And it is portfolio volatility that investors worry about, because it is their wealth volatility that drives a volatile consumption stream. The object of the imaginary or price-setting investor is stable consumption over time.

A volatile investment that nevertheless moves in a direction *counter-cyclical* to the majority of investments (a low covariance, low or negative beta asset) *is less risky* and will be valued at a premium, not a discount. Conversely, an investment that is highly correlated with the overall economy will be discounted even though its volatility (as measured by standard deviation) is low. It is this that leads to the divergence between actual and risk-neutral probabilities. An asset (perhaps a real estate investment) that is very volatile but with low or even counter-cyclical performance relative to overall market trends may sell at a relative premium even though the actual probability of the value increasing may be low.[9]

However, a (hypothetical) *risk-neutral* investor will value an asset by the mathematical *probability* of its value rising or falling and the *absolute* magnitude of the rise or fall. A *risk-averse* investor, which we assume represents the typical case, will value an asset by its rise or fall *relative* to an overall market portfolio.[10] We assume that market prices are determined by risk averse, not risk neutral, investors. That is why the share price $S = £20.00$ in the example above. The price reflects the buy/sell decisions of risk-averse investors calculating portfolio performance, not betting on the up/down probabilities of that particular share. It is this divergence that creates the opportunity for options, based upon risk-neutral arbitrage calculation, to perform as hedging instruments.

Second, we assumed a discount rate of 10 percent per annum. This requires some explanation. The option pricing mechanism sets up a risk-free hedge and the funds borrowed are, implicitly, therefore also available at the risk-free interest rate. This implies that the appropriate discount rate, *r,* in calculating option value is a *risk-free interest rate* such as, for example, a real (inflation-adjusted) sovereign long-term bond rate– see Eq (5.3). This is not the same as the discount rate applied to the call option itself. Option prices, like the underlying shares or assets themselves, can be very volatile. Therefore, they should be valued at the discount rate reflecting this underlying undiversifiable (or systemic) risk.

Third, the analysis here implicitly assumes transparent, liquid and zero-cost arbitrage markets. While this may be a fair approximation in many financial option markets, real estate markets seldom meet these conditions. Sundaram and Das discuss the problem of "replicability" in real-world situations. They cite the case of employee stock options and real options. [11]

> There is no easy "out" here. Depending on the particular situation, prices obtained via the standard techniques may still be useful as a benchmark. It may also be possible to modify the model to obtain a more appropriate price. Nonetheless, caution should be employed in interpreting the results too literally.

This does not invalidate the real options pricing analysis – but it does imply that in situations like these, the outcome of real option value calculations should be regarded as an information guide, not a precision instrument.

Consider a landlord's perspective

Now consider an example of option analysis with real estate applications.

Your tenant is about to make the final annual payment on a five-year lease. The rent, reflecting current market conditions, is £100 per annum, payable in advance. This seems an opportune time to negotiate a further five-year lease term.

Your tenant offers to renew the lease for a further five years at £100, the current rent. The future of market rents is uncertain. Rents might be higher or lower next year and into the future. No one can be sure.

How should you respond to the tenant's offer?

Consider the scenarios (Table 5.2):

The first payment on a new lease will be 12 months hence. Assume a discount rate of 10 percent per annum on the rental cash flow and the risk-free rate of interest is 4 percent per annum Therefore, the current ($t = 0$) PV of a roll-over of the existing lease of £100.00 for five years at a 10 percent per annum discount rate is £379.08 with the first payment due one year from now. You assess the outlook for $t = 1 \ldots 5$ as either a strong leasing market, in which (Up) case market rent will rise to £120 or a weak market (Down) with rent falling to £75.

For simplicity, we assume that a new lease will set the rent at a fixed level for five years. At the 10 percent discount rate PV = £454.89 for the Up case and £284.31 for the Down case.

As landlord, you risk losing the tenant at the termination of the current lease, in which case the space would be rented out at the prevailing market rent, which might be either £120 or £75. If we assume an equal (50 percent) probability for the Up and Down scenarios then

$$\text{NPV} = -£379.08 + \left(0.5 \times £454.89 + 0.5 \times £284.31\right) = -\ £9.48$$

Since NPV < 0 the NPV rule advises that the landlord should accept the tenant's offer to roll over the lease at the current market rent (£100) rather than take a chance on future market conditions.

But is this the optimal decision?

Table 5.2 Rental income – alternative scenarios

Net Present Value (£) – Discount Rate 10% p.a.

		< ----------	----------	Years	----------	---------- >
Scenarios	NPV(£)	1	2	3	4	5
Current	500.00	100	100	100	100	100
High case	600.00	120	120	120	120	120
Low case	375.00	75	75	75	75	75

Consider making the tenant a counter-offer: renew the current lease at £100 per annum but you, as the landlord, retain the option to reset the rent to the prevailing market rent (to be determined by an independent source) when the new lease commences one year from now.

Clearly, as the landlord, you will only enforce the mark-to-market option in the Up case, if market rent rises to £120.00.

Applying the option analysis

$$S = £379.08;\ u = £454.89 / £379.08 = 1.20;\ d = £284.31 / £379.08 = 0.75$$

In the Up case

$$Cu = £454.89 - £379.08 = £75.82$$

In the Down case

$$Cd = £0$$

Applying Eq (5.1A) and Eq (5.2A):

$$m = 2.25 \text{ and } B = £270.77$$

And the cash flow is in Table 5.3.

Applying Eq (5.1A) and Eq (5.2A), a perfect hedge involves purchasing the future lease (worth £379.08), borrowing £270.77 and shorting (or writing) 2.25 calls priced at £48.14 (2.25 x £48.14 = £108.31). In both the Up and Down cases the $t = 1$ payoff is £0.00. As the landlord, you are fully hedged.

As a cross-check, from Eq (5.2)

$$p = (1+r - d)/(u - d) = (1.10 - 0.75)/(1.20 - 0.75) = 0.56$$

and applying this probability

$$PV = 0.56 \times £454.89 + 0.48 \times £284.31 = £379.08$$

Table 5.3 Tenant lease – option arithmentic

Cash flow (£) Inflow (+)/Outflow (−)			
	$t = 0$	Payoff ($t = 1$)	
		Down	Up
Buy	−379.08	284.31	454.89
Borrow	270.77	−284.31	−284.31
Write calls	108.31	0.00	−170.59
Net cash flow	0.00	0.00	0.00

which is break-even (PV = £ 0.00) from the landlord's perspective.

The value of the option to the landlord is £108.31 and applying Eq (5.2) the additional value to the landlord of optionality in the lease contract is

$$\begin{aligned}\text{Additional value of flexibility} &= £108.31-\left(-\ £9.48\right)\\ &= £117.79\end{aligned}$$

Since the hedging strategy has a positive option value the better choice (from the landlord's perspective) is to stand firm on the current market rent in the new lease with the option to raise the rent to market. The tenant may be unwilling to accept a new lease which has only an upside option in the landlord's favour. But the landlord could offer an up-front cash incentive to the tenant of as much as £117.79 as a financial inducement to accept the optionality proposal.

Figure 5.1 illustrates the situation.

The long position payoff in Figure 5.1 illustrates the outcome in $t = 1$ relative to the $t = 0$ position if the landlord chooses to accept the market outcome. In the Up case the PV of the new lease is worth £454.89 and the payoff is

$$£454.89-£379.08 = £75.80$$

In the Down case the NPV = £284.31 and the payoff in the unhedged case is

$$£284.31-£379.08 = -\ £94.77$$

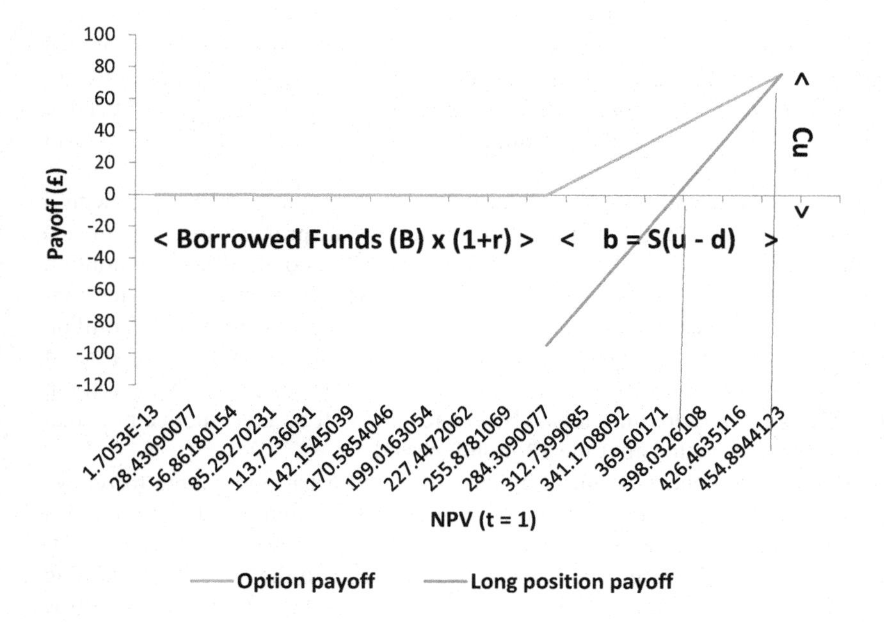

Figure 5.1 Option Strategy Payoff

Alternatively, if the tenant and landlord agree to the lease optionality condition, in the Down case the rent remains at £100.00 for the new lease and the landlord breaks even (£379.08), but gains

$$£454.89 - £379.08 = £75.82$$

in the Up case.

The effect of the constructed hedge is to create a call option payoff. Evidently, as Figure 5.1 illustrates, the funds borrowed are a function of the risk-free rate of interest (r) and the volatility of the anticipated Up and Down payoffs. The current value of this call option is £108.31, the balance of the funds after borrowing. The structure of the option, and its value, are determined by the volatility of the underlying asset as shown by the slope of the long position payoff.

Land value – more option arithmetic

A vacant building site in a CBD area has alternative uses and alternative development time frames. An ageing CBD office building is not so different from a vacant site. While the NPV of the future rental stream from the existing office building may be small, the value of the building may reside in its optionality value.

Importantly, as soon as a final decision is made, the option value attached to the site disappears. Since option values can be very high, this in itself can be a deterrent to making an irreversible decision. Valuable sites that sometimes lie undeveloped for many years bear witness to this fact.

Case study: An ageing inner-city 40,000 sqm office building delivers annual rent on short-term 12-month leases. The next annual rental payment of £250.00 per square meter per annum (psmpa) is due 12 months from now. It is expected that the rent will rise in the future at the rate of inflation, which is forecast to be 4.0 percent per annum in perpetuity.

The owner is considering demolition of the building and replacement with a building of identical size but with modern amenities. The project would commence 12 months from today, on the expiry of the current 12-month leases. Demolition cost is estimated to be £4 million and construction cost will be on a fixed price contract of £30 million payable in two instalments: the first instalment of £15 million is due on commencement of the project. The balance is due on completion, after 24 months. The owner has a local council Development Approval for the project, but a commitment is required immediately, and tenants will be given notice now that leases will be terminated after the 12-month period.

The council is contemplating a by-pass road and a parking garage to relieve congestion in the city centre. A decision will be made within the next 12 months. If the council project goes ahead inner-city amenities will improve and future office rent will be higher. The owner estimates that rents in the new office building will be £500.00 psmpa. on completion of the project and rents will thereafter grow at 2.0 percent per annum in excess of inflation. Otherwise, if the council project does not proceed, rents in the proposed new development are forecast to be £350 psmpa., growing at the rate of inflation. The owner estimates the probability (p) of the council project proceeding at 50 percent. The cost of capital applicable to the office market and the development project is 12 percent and the risk-free interest rate is 5.50 percent. The owner of the office building has the option of applying for a 12-month extension of the council's Development Approval, allowing the final decision on the project to be postponed.

Should the owner postpone the decision for 12 months or proceed with the development immediately? What is the current market value of the current and proposed office buildings and the land?

Table 5.4 sets out the cash flows for the alternative strategies – either develop now or postpone the development decision for 12 months.

Current asset value: The existing building delivers the next cash flow of £10.0 million in 12 months hence after which rent and therefore cash flow grows in line with inflation. Inflation (f) is forecast to be 4 percent per annum in perpetuity, a rate that is already embedded in the nominal discount rate, $d = 12\%$. We can therefore apply Eq (5.3).

$$\text{£}10.0\text{m}/(0.12 - 0.04) = \text{£}125.00\text{m}$$

to establish the PV of this asset.

Applying the method and notation described above, we can now assess the value of an option to delay the investment decision until next year (Table 5.5).

Project value: The development project, involves demolition and construction costs and future rental flows. Under the High scenario, the first rent (£505.00 psm) is received at the beginning of year 3 and this rent is forecast to grow at inflation (4 percent per annum) plus a real growth component of 2 percent per annum, or (1.04) x (1.02) – 1 = 6.08% p.a. Capitalising this at 12% – 6.08% by Eq (4.3), the value in Year 3 is £337.84 million. Under the Low scenario, the rent (£355.00) is capitalised at 12% – 5%, equal to £175.00 million. Applying a 50 percent probability to the Up and Down cases and discounting the sum (£256.42 million) to the present ($t = 0$) the proposed office building is valued at £182.51 million. Discounting the demolition and construction costs to the present, the PV of the completed project is £154.87. The PV rule, therefore, prescribes proceeding with the project immediately since £154.87 million > £125.00 million. But is this the optimal decision? This is the value excluding the option to postpone the project for 12 months.

Option value: Table 5.5 sets out the calculation of the additional value of the real option to postpone a decision on the project for 12 months.

Table 5.6 is the payoff matrix from the option strategy

Under the Up scenario, the asset is worth £212.83 million (taking account of demolition and construction costs), and £96.92 under the Down scenario. Hence $u = 1.70$ and $d = 0.78$. A hedged position requires a short position of 1.32 call options, priced at £25.11, or £33.13.

Table 5.4 Project financial analysis

Financial Parameters			
Discount rate (*d*)		12.0%	
Risk-free rate (*rf*)		5.5%	
Forecast inflation (*f*)		4.0%	p.a.
Real rental growth (g)	Project "Up" case	2.0%	p.a.
	Project "Down" case	0.0%	p.a.

				Year 0		1	2	3	4	--∞
Existing building	40,000	sqm								
Rent	Growth=inflation	£ psmpa				250.00	260.00	270.40	281.22	--∞
	x 40,000 sqm	£ mill p.a.				10.00	10.40	10.82	11.25	--∞
PV (Current)	Capitalised Rent	£ mill		125.00						
Project										
Proposed building	40,000	sqm								
Rent	Growth = inflation + real growth	£psmpa	Up					500.00	520.00	--∞
	Growth = inflation	£psmpa	Down					350.00	364.00	--∞
	x 40,000 sqm	£ mill	Up					20.00	20.80	--∞
	x 40,000 sqm	£ mill	Down					14.00	14.56	--∞
Cash Flow – Proposed Building										
	Demolition	£ mill				−4.00				
	Construction	£ mill				−15.00		−15.00		
	Capitalised rent	£ mill	Up					337.84		
		£ mill	Down					175.00		
Net cash flow		£ mill	Up		0	−19.00	0.00	322.84		
		£ mill	Down		0	−19.00	0.00	160.00		
PV (Current)	Up	£ mill		212.83	0	−16.96	0.00	229.79		
	Down	£ mill		96.92	0	−16.96	0.00	113.88		
PV (Current)	Probability adjusted (50%:50%)	£ mill		154.87	0	−16.96	0.00	171.84		

As a cross-check, we calculate the risk-neutral probabilities (see Eq (5.2))

$$p = (1+r - d)/(u - d)$$
$$= 0.30$$

Hence $1 - p = 0.70$

Table 5.5 Option value calculation

Real option analysis			
S	£125.00		
Su	£212.83	*u*	1.70
Sd	£96.92	*d*	0.78
Cu	£87.83	Cd	£0.00
m	1.32	B	−£91.87
mC = S+B	£33.13	*C* = mC/m	£25.11

Table 5.6 Option payoff scenarios (£ million)

Real option analysis			
S	£125.00		
Su	£212.83	*u*	1.70
Sd	£96.92	*d*	0.78
Cu	£87.83	Cd	£0.00
m	1.32	B	−£91.87
mC = S+B	£33.13	*C* = mC/m	£25.11

Applying these probabilities to the PV calculation (Eq (5.1))

$$-125.00 + ((0.30 \times £212.83) + (0.70 \times £86.82))/1.055$$
$$= £0.0$$

If the decision can be delayed for 12 months, the value of this option is £33.13. The project *including the real option* is therefore valued at

$$£154.87 \text{ million} + £33.13 \text{ million} = £188.00 \text{ million}$$

The optimal strategy is therefore to delay the final decision.

Land value: Since the land value is the residual (after demolition and construction costs) the land value is £154.87 million without the option. If the option is available and included, the land value rises to £188.00 million.

It is evident why developers often have a strong financial interest in persuading local authorities to adopt flexibility in allocating development rights. Local authorities may not fully understand the financial implications – and the value – of this flexibility. If they did they might adopt a tougher negotiating stance in discussions with prospective developers and their financial supporters.

Appendix 5A – Call and put options – a graphic approach

Options can take many different forms, limited only by the imagination of the contracting parties. Therefore determining the financial value of an option can be equally challenging. However, a good starting point is to establish the basic principles of the basic building blocks of option analysis – *call* and *put* options.

A *call* option is a contract that gives the holder the right (but not the obligation) to buy an asset at a predetermined price (the *exercise price*) at or before a specified date.

A *put* option is a contract that gives the holder the right (but not the obligation) to sell an asset at a predetermined price (the *exercise price*) at or before a specified date.

An American option allows the holder of the option to buy (or sell) the option at any time before or on the expiration date.

A European option can be exercised only on the expiration date.

Since the holder of an American option has greater flexibility than a European option, an American option will always be worth more than a European option with the same exercise price and expiration date.

The standard option pricing formula, the Black-Scholes equation, applies only to European options.

Consider a call option. The current price of the underlying asset (the Share Price) is S = £5.00. The call option gives the holder the right (but not the obligation) to purchase one share at the current price three months from now. Therefore the Exercise Price X = £5.00. We represent this situation in Figure 5.1A

SS represents the underlying share. The current price is S = £5.00. A rise in the share price to £6.00 implies a gain of £1.00. A fall to £4.00, a loss of £1.00.

CC represents a Call Option with an exercise price of X = £5.00. We assume that the price (or cost) of the option is £1.50. The option gives the holder the right to purchase the share for £5.00 on the expiry date. Therefore if, on the expiry date,

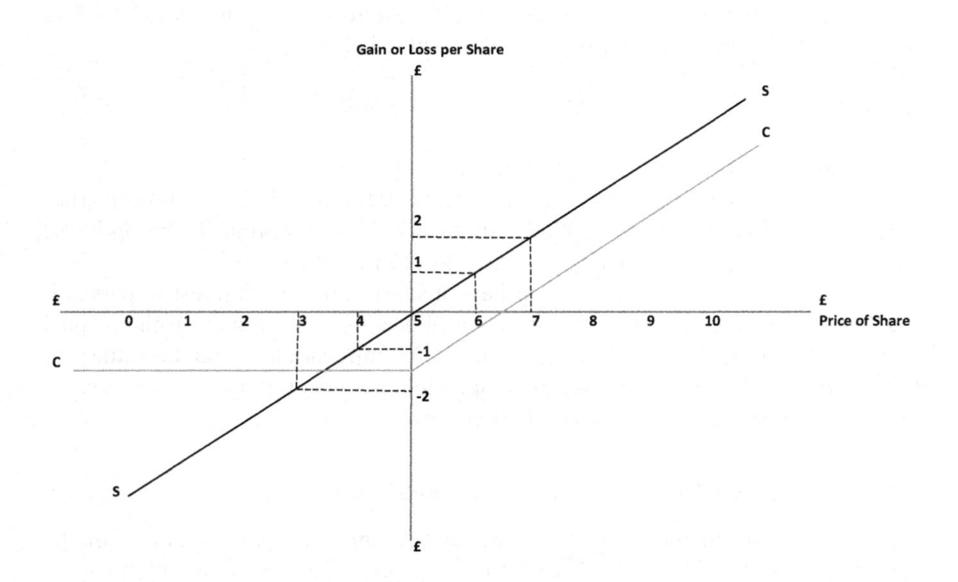

Figure 5.1A The Payoff Profile for a Call Option

S = £5.00 the option can be exercised but the option holder will have lost £1.50, the purchase price of the option. If on expiry date S = £6.50 the option holder breaks even (£6.50 − £1.50 = £5.00.).

Suppose on expiry date S = £3.00. The holder of the share has lost £2.00. The holder of the option will, of course not exercise the option – for the option holder the maximum loss is £1.50, the option purchase price.

Evidently, the call option will only be exercised if on the expiry date $S \geq$ £5.00. Therefore from the perspective of the seller (or writer) of the call option, the payoff is the inverse of the payoff to the option holder. Suppose the shareowner has sold (or written) the call option. Holding a share and writing a call option against the share is known as a *covered call* (shown as CC in Figure 5.1B).

If, on the expiry date, S = £5.00, the shareholder's value is S = £5.00 *plus* the proceeds from the sale of the call option, £1.50, and the shareholder/call writer is, therefore, £1.50 better off. However if on expiry date S = £6.50 the shareholder/call writer must deliver the share for £5.00 to the holder of the call. However, the addition of the value of the call sold (£1.50) establishes a break-even outcome. By writing the call the shareholder puts a ceiling of £1.50 on the future gains from holding the share, but in the event the S falls below £5.00, future losses are reduced by the income from the sale of the call (£1.50). The payoff from a covered call is illustrated by CC in Figure 5.1B.

Even with this simple graphical toolbox a wide variety of outcomes can be constructed for put options (the right to sell but not the obligation to sell – not buy) a share, and combinations of call and put options as well as outcomes when S ≠ X, as we have assumed in Figures 5.1A and 5.1B.

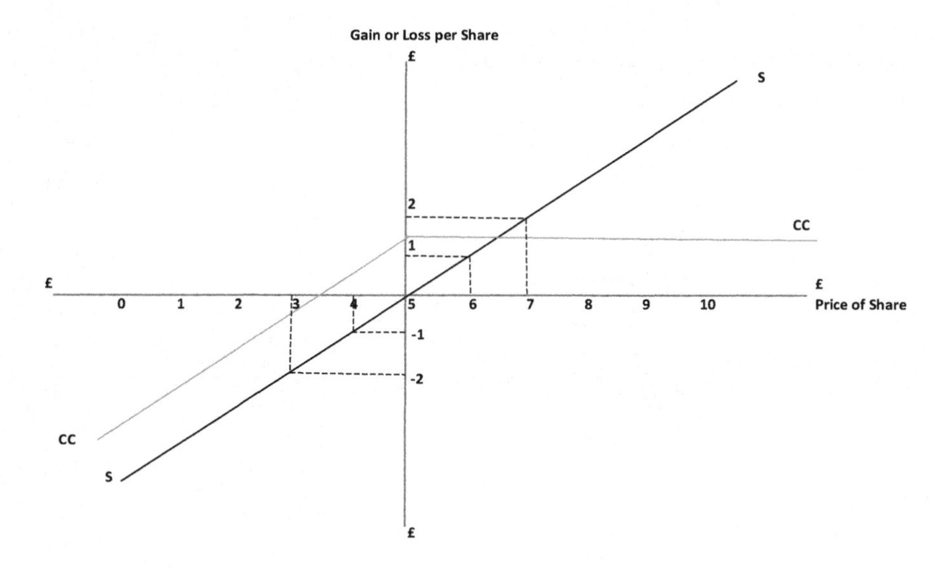

Figure 5.1B The Payoff Profile of a Covered Call Option

Appendix 5B – The basic formulas – how did we derive the Table 5.1 numbers?

The formulae for the example above are derived as follows:

Define S as the current share price, B as the borrowed funds and rf as the risk-free interest rate. We use the risk-free rate because the pricing of the call option is determined by a risk-free arbitrage strategy. We set up a portfolio comprising one share, S, borrowed funds, B and write (sell short) a number of call options, m. At $t = 1$ we repay the borrowed funds plus the interest cost, $(1 + rf)$B and, in the Up case we purchase the m options that were sold short at $t = 0$. In the Down case, the option expires worthless

Under Up and Down alternative scenarios the outcomes at $t = 1$ are:

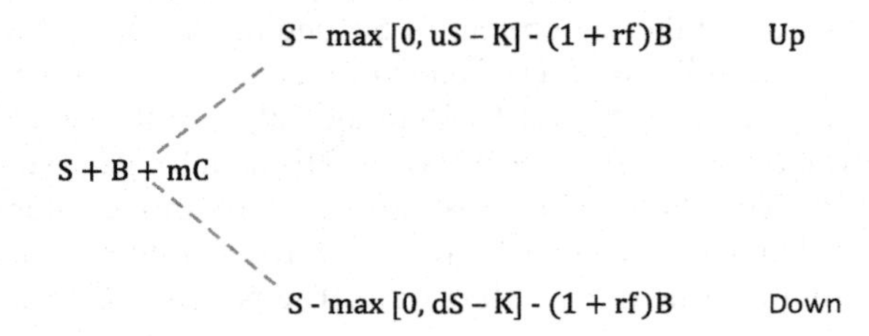

Since the investor is fully hedged, the strategy is risk-free and the outcome in $t = 1$ for the Up and Down cases must be identical:

$$uS - (1+rf)B - mCu = 0$$
$$dS - (1+rf)B - mCd = 0$$

where m is the number of call options sold.

Solving these equations:

$$m = \frac{S(u-d)}{Cu - Cd} \quad (5.1A)$$

$$B = \frac{S(uCd - dCu)}{(Cu - Cd)(1+rf)} \quad (5.2A)$$

And at $t = 0$

$$mC = S + B$$

or

$$C = (1/m)(S + B)$$

From Table 5.1 where $rf = 10\%$, $S = £20$, $u = 1.20$, $d = 0.67$ and $B = -£12.18$, it follows that $m = 5.30$ calls $= £7.82$ so $C = £1.48$.

Note that if Cd = 0 the equations can be further simplified to

$$m = \frac{S(u-d)}{Cu}$$

and

$$B = \frac{-dS}{(1+rf)}$$

Figure 5.1 shows this in graphical form.

Notes

1 Savills, *The 10 Most Valuable Real Estate Markets in the World*, https://www.savills.com/blog/article/219340/international-property/the-10-most-valuable-real-estate-markets-in-the-world.aspx.
2 The Global Economy.com, https://www.theglobaleconomy.com/rankings/stock_market_turnover_ratio.
3 See for example Sick, G., and Gamba, A., Some Important Issues Involving Real Options: An Overview, *Multinational Financial Journal*, 14, nos. 1/2 (2010), pp. 73–123.
4 This is a financial option, but the principles apply to real options as well as will be demonstrated. The example is derived from J. C. Cox, Option Pricing: A Simplified Approach, *Journal of Financial Economics*, 7 (1979), pp. 229–263.
5 Appendix 5A explains the basic features of call and put options.
6 Appendix 5B shows the mathematical derivation of the values in Table 6.1.
7 This is "selling short" that is, selling for payment now, but with delivery on the maturity date ($t = 1$).
8 See Chapter 10 for a more detailed discussion of asset pricing.
9 Cochrane, J. H., *Asset Pricing* (University of Chicago, 2000) (especially Chapters 1 and 2) provides an accessible, though mathematically rigorous, discussion of the fundamental drivers of asset and option pricing.
10 Cochrane (2000), p. 23.
11 Sundaram, R.K., and Das, S. R., *Derivatives: Principles and Practice*, 2nd ed. (New York: McGraw Hill, 2016), p. 239.

6 Change the rules, change the game

How property rights regulate outcomes

The laws and conventions that define property ownership, occupancy and exchange are a key differentiator between prosperous and poor societies. Some property ownership practices and legal arrangements encourage high levels of economic efficiency and wealth creation; other arrangements, less so.

Establishing, maintaining and enforcing private property rights as well as the limits to these rights is costly; the absence of clear definitions of these rights can be even more costly. Once established, property ownership and access rights can become closely aligned with personal wealth, business and personal financial commitments and social status. Therefore, these rights can be very difficult or costly to change even when circumstances alter. Preserving existing property rights and arrangements when circumstances change can also deliver sub-optimal social and financial outcomes.

This chapter describes how alternative property ownership arrangements can arise and examines how different property rights arrangements can lead to diverse outcomes. As a bonus, we will also illustrate the concept of *economic efficiency*. The two subsequent chapters, Chapters 7 and 8, examine what strategies are available when property rights "fail" and offer insights into policies to overcome barriers to establish more efficient and equitable property rights arrangements.

The tragedy of the commons

The Dutch explorer, Captain Jacob Roggeveen, and his crew were the first Europeans to visit Easter Island in the South Pacific. When they arrived, on Easter Monday, 5 April 1722, they encountered a society in disarray. Roggeveen reported no native land animals on the island larger than insects, and no domestic animals larger than chickens. The people of the island, estimated to number around 1,500–3,000, were on the verge of starvation. The inhabitants depended on fish, birds and sweet potatoes for a hand-to-mouth existence; cannibalism was not unknown.

As Jared Diamond recounts, the English explorer, Captain James Cook, who visited Easter Island in 1774, described the islanders as "small, lean, timid and miserable."[1]

DOI: 10.1201/9781003111931-6

But the surprising fact – revealed by subsequent archaeological research – was that the islanders had not always lived like this. In the not-so-distant past Easter Island had been host to a population of around 17,500. Jared Diamond suggests that the population of Easter Island declined by around 70 percent from a peak sometime between 1400 and 1600 and the arrival of the early European navigators in the 1700s.

Why?

We shall return to Easter Island. But let's first take a step back.

Historically property ownership arrangements often emerge as a response to new opportunities or emerging threats.

For example, if population numbers rise in an agrarian society, arable and grazing land becomes scarce. The result may be conflict. An alternative is the formalisation of arrangements for defining, managing and exchanging usage rights and parcels of land. Economists with an interest in property rights often commence their analysis with a parable that goes something like this:

> Two tribes live in a forest. Game and fruit are plentiful and the tribes both prosper. The tribes encounter each other occasionally in their wanderings but have no other contact. Over time, however, as both tribes increase in numbers, game and fruit becomes harder to find. The tribes find they are competing for the same resources that are now becoming scarce. A summer drought brings matters to a head.
>
> The wise leaders of the tribes arrange a meeting in a clearing on the banks of the river that runs through the middle of the forest. They agree that one tribe will confine its activities to the north side of the forest and the other tribe will occupy the south side. The boundary is conveniently defined by the river.
>
> As it happens, game is more prolific on the north side of the forest and fruit on the south side. But this is not a major problem because the tribes agree to meet each full moon in the clearing by the river to barter fruit and game.

But a range of other outcomes can be imagined. As an alternative to communal rights, some societies invent the concept of individual property ownership or tenure. For example, in Western Europe, the medieval feudal system developed as a complex system of leasehold and freehold ownership rights. Reciprocal obligations were devised to support agricultural production and at the same time to provide funding and resources for defence against foreign invaders as well as the provision of some public services such as roads, forest clearing and drainage. The church was engaged to provide healthcare and poverty relief.

No one sat down and planned feudalism as an international system. It emerged from necessity and mutual advantage, with the usual human admixture of compulsion, greed, faith and fear. Built around agricultural production and military protection, paid for by systems of taxation based upon labour hours, levies on agricultural produce and compulsory military service to the local chieftain, the feudal system provided a pragmatic basis for organising a society in the Middle Ages.

But slowly over time, the complex system collapsed.

Communications improved; technology advanced; the Black Death, which might have erupted in Asia and was spread to Europe by trading ships in the 14th century, caused a sharp fall in population and therefore a scarcity of labour. As a result, wages rose. Serfs found that they could bargain their labour with the neighbouring chieftain. The feudal arrangements that previously had bound people to the land for their own protection now impeded their economic progress.

As evidence of the persistence of ancient property rights, residual features of the feudal system are still visible today in the laws and property ownership arrangements of the United Kingdom and many European countries, even though the original drivers of the system have generally vanished (though the greed, of course, remains). The feudal system collapsed, but its departure was a gradual process, spread over centuries.

Sometimes we do not have the luxury of time. And abrupt changes to laws, particularly property laws such as zoning regulations, create winners and losers.[2]

In North America, in contrast to Europe, private property rights emerged as a clear and explicit policy objective. It was the prominent Virginian landowner and future President of the United States, Thomas Jefferson, who proposed in 1774:

> ...each individual of the society may appropriate to himself such lands as he finds vacant and occupancy shall give him title.[3]

Jefferson proposed this policy as a sharp contrast to what he saw as the European feudal concept that in principle all land belonged, ultimately, to the king.

Taking Jefferson's lead, on 20th May 1785 the United States Congress passed an Ordinance to survey and sell the land west of the Ohio River. The process of defining property boundaries was carefully prescribed:

> The surveyors shall proceed to divide the said territory into townships of six miles square, by lines running due north and south and other crossing these at right angles...The lines shall be measured with a chain [22 yards];[4]

And on 30th September 1785 Thomas Hutchins, the first Geographer of the United States, set out from the Point of Beginning near Pittsburgh, with his chains (22 feet in length) to triangulate America. His template gave rise to the rectangular farm boundaries and the street grids of cities that characterise vast areas of the continental United States from coast to coast today, their dimensions often defined (to the surprise of current owners) in multiples of an English cricket pitch.

Property rights, however, do not always emerge as an efficient outcome to a land allocation problem, as the Easter Island example illustrates. The earlier vibrancy of that society is attested by the existence of the large stone statues, some weighing up to 270 tonnes, for which Easter Island is famous. Whatever the religious or social motives for carving, transporting and erecting these huge statues (and this remains a mystery) there can be no doubt that they represent a massive commitment of technical skills and labour resources for a pre-industrial society. To compound the mystery, the task of moving the statues from a quarry to the place of erection required a complex system of ropes (made from fibrous tree bark) and rolling logs. However, not a single tree taller than 10 feet was to be found on Easter Island in the 1700s. The elimination of trees on Easter Island had many adverse consequences, not least a decline in the number of nesting birds and erosion of the soil.

So, what had happened?

We cannot now be sure, but Easter Island does look like a real-world example of the so-called Tragedy of the Commons. First discussed in an early pamphlet,[5] and later popularised in a widely referenced article in 1968,[6] it is a parable along the following lines:

> You live in an isolated village. Each villager has a small private plot of growing crops but the domestic animals, let's call them cows, are free to graze on the common pasture surrounding the village. Over time, the village population increases and so, too, does the number of cows. Grazing land becomes scarce.
>
> A logical response is to limit the number of cows and perhaps initiate a policy to distribute the milk and meat. But the villagers can't agree on a process for this. As a result, the cows continue to increase in number: but they get smaller and thinner. Milk and meat supplies dwindle. The response of each villager is to keep even more cows to maintain the family diet.
>
> The result is overgrazing, soil erosion, conflict and perhaps, eventually, the collapse of the society. At the least, the villagers' standard of living is lower than would be possible under an alternative regime that would limit the number of cows and apply some optimizing formula to allocate the limited grazing land.

So why did the Easter Island society fail to produce the win-win outcome that has emerged in at least some other societies? Why not convene a tribal meeting and thrash out the issues?

Jared Diamond has posed an important question about Easter Island:

> 'What did the Easter Islander who cut down the last palm tree say as he was doing it...like modern loggers did he say 'Jobs, not Trees? Or 'Technology will solve our problems, never fear we'll find a substitute for wood'?[7]

The Easter islander who cut down the last palm tree may have done so out of ignorance. But it is just as likely that he understood the consequences of his actions. He may also have understood that if he did *not* cut down the last tree (to provide winter firewood to warm his family) then some other islander *would* cut it down. So, the rational islander would cut down the last tree.

A fanciful parable?

Not if you visit parts of Africa, South America or Asia. Nor is this behaviour confined to primitive societies. Overgrazing in tribal communities arises from the same divergence between private and public incentives that leads to peak-hour traffic in London on weekdays, parking congestion in Snowdonia on August Bank Holiday and the undignified scramble in the few fair-weather days available to reach the summit of Mount Everest in May each year. In a world of scarcity, calculations of private costs and benefits from individual actions can and do diverge from costs and benefits more broadly defined for society as a whole.

The result may be overutilisation of grazing land, traffic congested roads and sometimes much worse outcomes. Importantly, there is no easy way for the individual to escape from this dilemma. Even if I decide to commute into town by train today as a noble contribution to easing road congestion, someone else, making the opposite decision, will probably offset my selfless action as I board the 8.35am to Euston. In choosing to drive my car, my choices and decisions are no different from the man with the axe on Easter Island.

It is not difficult to identify the Tragedy of the Commons as arising from a lethal combination of private ownership of domestic animals and communal ownership of the land – not unlike our current arrangements of private car ownership and communal ownership of the roads – with familiar consequences of delays, congestion and inefficiency.

The economist Harold Demsetz (1930–2019) has offered an explanation for the emergence of property rights:

> ...property rights develop to internalize externalities when the gains of internalization become larger than the cost of internalization. Increased internalization, in the main, results from changes in economic values, changes which stem from the development of new technology and the opening of new markets, changes to which old property rights are poorly attuned...A broad range of examples can be cited that are consistent with it: the development of air rights, renters' rights, rules of liability in motor accidents, etc.[8]

We cannot be sure why this process seems to have failed on Easter Island. But we also know that across the broad perspective of history it is poverty, not affluence, that most commonly defines the human condition. Devising and enforcing laws is costly. But so, too, is the absence of laws. It is a truism (albeit an insightful one, as Demsetz shows) that laws, including laws that define property rights, will emerge when the anticipated communal benefits from these laws exceed the communal costs. Further, establishing a legal system, even a system of efficient laws, will not of itself guarantee efficient outcomes. Once established, the legal system must be effectively and efficiently enforced. And amended as technical, social and demographic factors change. This is also costly.

Planning a residential development project

Different property ownership arrangements, even in a modern context, can lead to very different outcomes, even for apparently rational investors and policymakers.

Table 6.1 A blueprint for a residential development project

		Total	*Total*	*Total*	*Per Unit*	*Marginal*	*Marginal*	*Stamp*
Apartment	*Price*	*Revenue*	*Cost*	*Profit*	*Profit*	*Revenue*	*Cost*	*Duty (10%)*
Number	-- *£'000* --							
	(1)	*(2)*	*(3)*	*(4)*	*(5)*	*(6)*	*(7)*	*(8)*
1	6.6	6.6	2.3	4.3	4.30	6.6	2.3	0.66
2	6.3	12.6	3.5	9.1	4.55	6.0	1.2	1.26
3	6.0	18.0	4.2	13.8	**4.60**	5.4	0.7	1.80
4	5.6	22.4	5.6	16.8	4.20	4.4	1.4	2.24
5	5.2	26.0	7.6	**18.4**	3.68	3.6	2.0	2.60
6	4.7	28.2	10.5	17.7	2.95	2.2	2.9	2.82
7	4.3	30.1	13.9	16.2	2.31	1.9	3.4	3.01
8	3.9	31.2	18.7	12.5	1.56	1.1	4.8	3.12
9	3.5	31.5	25.3	**6.2**	0.69	0.3	6.6	**3.15**
10	3.1	31.0	34.1	–3.1	–0.31	–0.5	8.8	3.10

The BOLD cells shows the key metrics under alternative property-rights arrangements

Suppose you are contemplating investment in a residential apartment development on a block of vacant land. You hire a multidisciplinary team of architects, quantity surveyors and real estate agents to prepare a feasibility report demonstrating the financial outcomes across a range of alternatives ranging from low to high-density developments. The consultants provide the information in Table 6.1 below:

As Table 6.1 shows, you can, if you choose, construct one large, palatial, apartment or residence which would have an estimated market value of £6,600 (Col (1), the money values are arbitrary). The cost of this project would be £2,300 (Col (3)), leaving a profit of £4,300 (Col (4)).

An alternative would be to construct two apartments. Because they are smaller the market value of each apartment will be lower, £6,300, and total revenue will be 2 × £6,300 = £12,600 (Col (2)). Your construction cost will rise to £3,500, leaving a larger profit of £9,100 (Col (4)).

Note that the cost rises with the second apartment, but not proportionately. There are scale benefits from constructing two dwellings instead of one. The average cost per apartment falls from £2,300 to £1,750. The incremental revenue from erecting the second apartment Marginal Revenue (MR) is £12,600 – £6,600 = £6,000 (Col (6)) and the incremental cost Marginal Cost (MC) is £3,500 – £2,300 = £1,200 (Col (7)).

So, by choosing to construct two apartments instead of one, the total profit has increased from £4,300 to

£12,600 – £3,500 = £9,100

Given the information in Table 6.1, how many apartments should you construct?

Unfortunately, the consultants cannot tell you that.

You need a real estate analyst, who will explain that the answer to your question depends on the ownership arrangements attached to the project.

- **Commercial property developer**: If you are a commercial property developer seeking to maximise profits through the development and sale of these apartments, the optimal number of apartments is five, generating a profit of £18,400 (Col (4)). Notice that the incremental revenue from adding the fifth apartment is £3,600 (Col (6)) and the incremental cost is £2,000 (Col (7)). Had a sixth apartment been erected the addition to revenue would have been £2,200 but costs would have increased by £2,900. Profit would have declined by £700. The additional resources absorbed by the sixth apartment would exceed the financial benefits generated.
- **Syndicate**: Suppose a community group owns the land and decides to develop the project on the basis that each syndicate member will own one of the apartments. How many apartments should the syndicate construct? If five apartments are constructed each participant earns a profit of £18,400 ÷ 5 = £3,680 (Col (5)). However, if the number of apartments is limited to three, each participant earns £4,600. Clearly, a syndicate will limit the number of apartments to fewer than the commercial developer will choose. Syndicates (such as legal and medical professional bodies and trade unions) often act to limit entry to their organisations to maximise benefits for existing members. Note that the market value of the fourth apartment (£4,400) exceeds the construction cost (£1,400) but this will not be attractive to the syndicate.
- **Local Council**: The local council collects stamp duty at the rate of 10 percent of the sale price of each new residence. As mayor, one of your responsibilities is to maximise the revenue of the council that will fund the provision of social services and amenities to the wider community. Your choice must therefore be a development of nine apartments, yielding a stamp duty of £3,150 (Col (8)). Local councils, unsurprisingly, are often in favour of high-density living, sometimes explained as a way of protecting the environment and reducing "urban sprawl."
- **Break-even**: You are in charge of a not-for-profit housing commune and have been allocated this land to "optimise housing benefits to the community." You will choose to erect nine apartments, which is the maximum number that can be constructed: the tenth apartment will drive you into a loss of £3,100 (Col (4)). Note that although you make a small profit on the ninth apartment, the additional cost of erecting it is £6,600 (Col (7)) while revenue increases by only £300. Not only has the ninth apartment absorbed more resources in its construction than it contributed to market value. The addition of this ninth

apartment has driven down the value of the other eight apartments by £400 each (Col (1)). Both the Local Council outcome and the Break-Even strategy result in an *economically inefficient* outcome, ignoring any social or community benefits that may indeed exist, but are not reflected in the financial data set out in Table 6.1.

- **Government control**: Suppose the government owns the project. Government behaviour is hard to predict because the objectives are often complex and frequently obscure. But a reasonable set of alternatives would be:
  - **Benign dictatorship**: A well-intentioned government will either erect and sell five apartments, making a profit of £18,400 or put the project out to tender to the highest bidder. In a competitive tender process, the highest bidder will bid £18,400 and erect five apartments. The benign government will then use the proceeds from the sale for other, perhaps desirable, community purposes or reduce taxes on the community.
  - **Corrupt dictatorship**: A government, or its agents, might conspire with a developer to share the proceeds of a development. In this case, the option of five apartments might still be the most attractive since this maximises the surplus funds available for distribution to the conspirators.
  - **Vote-buying democracy**: A government anxious to maximise its popularity at the ballot box will be likely to choose a larger number of apartments, particularly if the project is located in a marginal voting constituency. A ten-apartment complex might be selected, even though the tenth apartment reduces total revenue and drives profit into negative territory. Of course, conversely, a high-density project might encounter opposition from existing residents in the area. Since existing voters probably carry more weight than prospective voters in the proposed development it might even be politically desirable to abandon the project entirely, despite the evident financial benefits.
  - **NIMBY or Green government**: A democratically elected local council, or one dependent on green votes, might choose to abandon the whole scheme in the interests of protecting the environment and the interests of the current local residents. Alternatively, a high-density project might be preferred to limit the impact of population growth and urban sprawl across the broader environment, perhaps protecting the interests of existing residents in the adjacent green belt.

The key conclusion from this analysis is that while we can define the *economically efficient* outcome of this project, the actual outcome will vary depending on the ownership arrangements and therefore the incentive structure. When property ownership rights change, so too do choices.

Economic Efficiency: Economists traditionally judge policy outcomes according to the two criteria of *efficiency* and *equity* (or "fairness"). The definition of *equity* is a subject for debate and there is no generally accepted definition, but the definition of *efficiency* is widely accepted.

In Table 6.1 the *efficient* outcome is five apartments. This is because the additional value of the fifth apartment (MR = £3,600, Col (6)) exceeds the additional cost of producing it (MC = £2,000, Col (7)). For the sixth apartment cost exceeds revenue.

Economic efficiency requires that production expands as long as MR exceeds MC.

Maximum efficiency is defined by the condition that output expands to the point where MR converges to MC from above.

This lies between five and six apartments (Table 1). Beyond the point where MR = MC, the resources absorbed in the production of each additional apartment exceeds the value produced. This is an inefficient outcome.

Note that in Table 1 the Commercial Developer will maximise the efficiency of the project. But so, too, might a Benign Dictator, or even a Corrupt Dictator.

Applying this definition of economic efficiency, Table 6.1 also shows that the Break-Even strategy of building nine apartments is inefficient. The additional four apartments add £31,500 − £26,000 = £5,500 of value (Col (2)) at a cost of £25,300 − £7,600 = £17,700 (Col (3)). The resources absorbed by the construction of an additional four apartments would have delivered greater value if allocate elsewhere, say to an alternative residential project.[9] And note that there are a variety of alternative ownership arrangements that, at least in theory, can lead to an efficient allocation of resources.

Conclusion – the bottom line

Property rights typically emerge and alter as a response to new social and economic circumstances. The opportunities and obligations associated with property ownership and occupancy in turn create incentives that influence human choices and behaviour. Even casual observation reveals that efficient systems for defining, adjusting and enforcing property rights are rare. Poverty, not affluence, is a global norm. Conflict, either political or physical, is not uncommon.

While economic efficiency is one of the target objectives of any public policy it does not follow that it is the only objective. The two criteria economists use to judge alternative policies are *efficiency* and *equity*. The definition of *efficiency* is a technical matter that is generally agreed. The criterion of *equity* (sometimes described as "fairness") is however susceptible to multiple different interpretations.

While it seems probable that a commercial developer and (perhaps) a benign dictator will meet the efficiency criterion in our property development example, it does not follow that this is the *only* desirable objective. Other considerations may also be important. The proposed development may have negative or positive spin-off impacts that Table 6.1 does not take into account. These spin-off impacts, which are not fully reflected in market prices and therefore do not directly influence the choices of the developer, are called "externalities." Examples might be the overshadowing of adjacent dwellings or increased traffic congestion. And as another challenge, when circumstances, infrastructure or technology change, laws and regulations must often adapt.

We shall examine *externalities* and the related concept of *market failure* in the next two chapters. We shall also explore the problems that arise when property rights change, imposing windfall gains or unanticipated losses, and in Chapter 7 we shall consider potential strategies to facilitate changes to laws and rules, avoiding the zero- or negative-sum outcomes that often precipitate sub-optimal outcomes and sometimes ignite conflict.

Notes

1 Diamond, J. M., *Collapse: How Societies Choose to Fail or Survive* (London: Penguin, 2005), p. 109.
2 We discuss this problem in more detail in Chapter 7.
3 Linklater, A., *Measuring America* (London: HarperCollins, 2003), p. 50.
4 Linklater, 2002, p. 70.
5 Lloyd, F. W., *Two Lectures on the Checks to Population* (Oxford, England: Oxford University, 1833).
6 Hardin, G., The Tragedy of the Commons, *Science,* 162 (1968), pp. 1243–1248.
7 Diamond, 2005, p. 114.
8 Demsetz, H., Toward a Theory of Property Rights, *The American Economic Review,* 57, no. 2 (1967), pp. 347–359.
9 Note that the MR = MC criterion for economic efficiency is subject to qualifications in the case of, for example, renewable and exhaustible resources (see Chapter 9).

Bibliography

Demsetz, H. 1967. Toward a Theory of Property Rights. *The American Economic Review,* 57(2), pp. 347–359.
Diamond, J. M. 2005. *Collapse: How Societies Choose to Fail or Survive*. London: Penguin.
Hardin, G. 1968. The Tragedy of the Commons. *Science*, 162, pp. 1243–1248.
Linklater, A. 2003. *Measuring America*. London: HarperCollins.
Lloyd, F. W. 1833. *Two Lectures on the Checks to Population*. Oxford, England: Oxford University.

7 When property rights fail – compensation and rent-seeking

Getting to "fair value"

Ownership of property assets, including intangible assets such as licences and patents, confers a package of rights and obligations. Some rights are under the direct control of the owners: rights can often be sold, rented out or temporarily divested. However, rights can also be curtailed, modified or regulated by, for example, changes to zoning regulations or compulsory purchase. External factors such as changes to taxation or rating arrangements, environmental regulations or transport and social infrastructure (such as a local school or hospital) can indirectly influence the value and transferability of property rights.

Changes to property rights ("shifting the goalposts"), however desirable these changes may sometimes be on economic efficiency or social equity grounds, can impose substantial losses or gains on property owners. This is particularly important when changes are unanticipated or retrospective. The mere possibility of future adverse rule changes may deter prospective investors or at least lead to elevated investment hurdle rates. The economic cost of investment or market activity foregone is difficult to measure, but potentially very large indeed. The concept of "market value," often used in the assessment of compensation payments in the event of adverse changes, is a convenient but not necessarily an equitable or "fair" basis for assessment. Opposition from the prospective losers in these transactions is one reason why implementing change is often difficult. But changes in the rules of the game that lead to increased efficiency can in principle create a "win-win" outcome for all parties. This chapter explores the economic underpinnings of compensation and how "shifting the goalposts" can be facilitated, even welcomed by the affected parties.

While discussion of compensation tends to focus on losses, property rights changes can also deliver windfall gains. The prospect of windfall gains can stimulate strategic behaviour to secure these rights. This activity, which redistributes but does not create wealth, is sometimes termed "rent-seeking." Rent-seeking behaviour, like externalities (discussed in Chapter 8) is also pervasive. The economic cost of resources allocated to rent-seeking, like externalities, is also difficult to measure but potentially is very large in the real estate sector.

DOI: 10.1201/9781003111931-7

Compensation – is it ever enough?

Walk down a suburban street. It's not uncommon to see a For Sale sign displayed on one or two of the houses. Most homes on the street, however, are not advertised for sale. This is not an oversight by the owners. It's a reasonable assumption that these owners value their houses more highly than a likely cash offer, equivalent to the prevailing market value of their home. At some higher price, probably, most houses in the street would be willingly offered for sale by their owners. But for at least some owners, with, say, a strong affiliation with that location, the price that would induce them willing to sell may be very high indeed.

Let's introduce two concepts:

Willingness to pay (WTP): The amount I am prepared *to pay voluntarily* to acquire an asset, property, right or service.

Willingness to accept compensation (WTAC): The amount I am willing *to accept in voluntary compensation* for relinquishing or being deprived of an asset, property, right or service.

When I visit the local supermarket, if my favourite brand of breakfast cereal is out of stock I am probably willing to settle for a similar brand as an alternative or in *compensation.* I am disappointed, but it will not spoil my whole day. In the ordinary course of life, WTP for an item or service is typically very close to WTAC in the absence of the preferred brand or item. This is especially the case when there are likely to be close substitutes, as with competing brands of breakfast cereal or toothpaste.

Example: The houses for sale in the suburban street under normal market conditions transact at market prices. When a transaction is concluded between a willing buyer and a willing seller we know can conclude that the purchaser's WTP is *at least as great as* the agreed price; the vendor's WTAC is *no less than* the agreed price.

In the case of a voluntary sale, at the agreed price

$$\mathrm{WTP}(p) \geq \mathrm{WTAC}(v)$$

where p denotes purchaser and v denotes vendor.

We also have good grounds for believing that in the case of the neighbouring houses not currently offered for sale:

$$\mathrm{WTP}(p) < \mathrm{WTAC}(v)$$

Compulsory acquisition of these homes at assessed or prevailing market prices, say, to build a motorway, would likely represent a net financial or psychic loss to the owners.[1] We can anticipate opposition from current owners even when offered a "fair" or "market" price for their homes, as defined by the recent voluntary sale of a similar house in their street. To understand why the current owners oppose the compulsory acquisition, and to consider what can be done about it in property markets (and in life generally) we introduce the concept of *consumer surplus*.

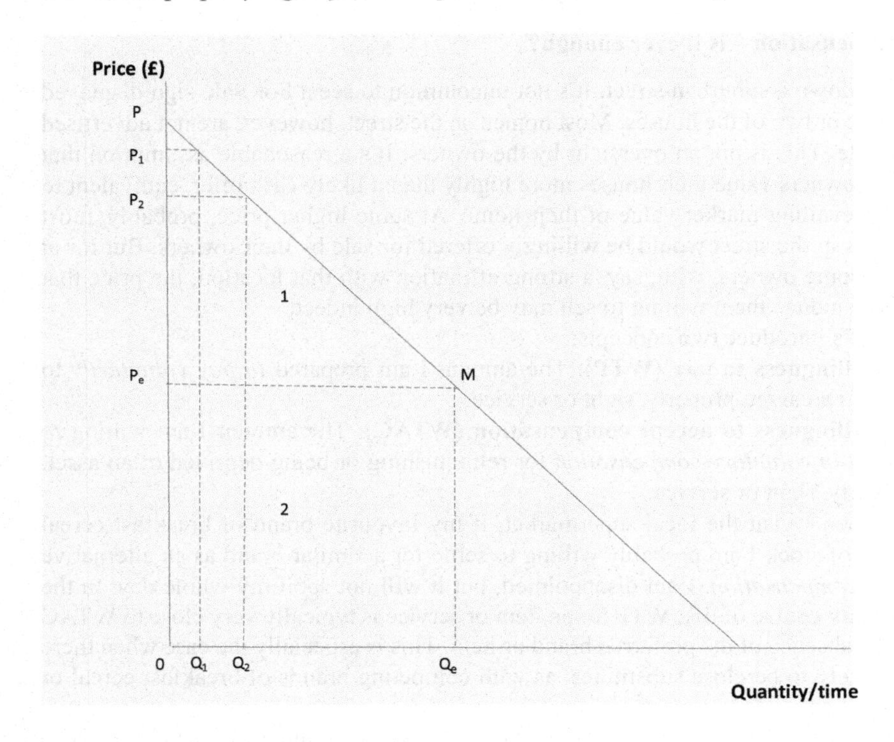

Figure 7.1 The Demand for Widgets

Figure 7.1 shows an orthodox demand line.

Note that the horizontal axis shows the *level of demand* (quantity demanded per time).[2] Let's assume the market price of this item (widgets) is P_e. Define the period as one month. At the price P_e the quantity of widgets demanded per month is Q_e. The demand line also shows that if the quantity available for purchase in a given month were only Q_1 the consumer would be willing to pay (WTP) a price of OP_1 per unit for this very limited availability. Having paid P_1 for Q_1, if an additional small quantity $Q_2 - Q_1$ were then to become available unexpectedly, the consumer would be WTP P_2 for the additional quantity.[3]

However, the prevailing market price is actually P_e. Therefore, for the quantity Q_1 the consumer is paying a price of P_e for an item that we know he or she values at P_1. Similarly, for the additional quantity $Q_2 - Q_1$ the value P_2 exceeds the actual market price P_e. If the market price is P_e for all widgets then, for all widgets up to the Q_eth widget the consumer is paying less than the value he or she attaches to each widget of the prior $Q - 1$ or the intra-marginal widgets. In a sense therefore the consumer is getting "something for nothing." This is known as "consumer surplus" and summing all the Q_1, Q_2, ... units together up to Q_e gives Area 1 (P_ePM). The total benefit (or "utility") enjoyed by the consumer is Area 1 *plus* Area 2 (OP_eMQ_e). This represents the maximum amount the consumer would be WTP if the vendor could negotiate the sale of each widget individually and if the consumers were blindly ignorant of the fact that after each transaction (Q_1, Q_2 ... etc.) yet one more widget would be offered for sale.[4] Of course, the creation of artificial or temporary scarcities to elevate prices is a popular retailer's tactic: "while stocks last," "offer must end Tuesday," "never to be repeated," etc.

This has been an interesting (depending on your point of view) thought experiment. Now let's reverse the process.

If we now confiscate all Q_e widgets and compensate the consumer at the market price, P_e, the consumer receives only Area 2 or OP_eMQ_e. The additional psychic (or utility) loss to the consumer is Area 1 or P_ePM. Compensation at the market price is an adequate refund for the consumer's financial outlay, but this does not adequately compensate the consumer for the utility loss of the confiscated widgets.

There is a little more to the story, however.

- **First**, this explanation has cut a few technical corners in the interest of explaining the concept of *consumer surplus*. Implicit in the explanation is the assumption that when the price of widgets changes, all other factors remain unchanged – the economist's *ceteris paribus* assumption. But a fall in the price of widgets must mean that the consumer's real income – what's left over after purchasing Q_e widgets – is higher than it was before. So, the *ceteris paribus* assumption has been violated. Technically the demand line in Figure 7.1 is a so-called Marshallian demand line, which incorporates *both* a relative price (or substitution) effect and an income effect – so it technically overstates the magnitude of the consumer surplus, although the principle remains clear.
- **Second**, suppose the consumer, having received Area 2 in cash, is able to purchase identical, or very similar widgets from some other source. In this case, Area 1 is completely, or largely, recovered. A working rule is that if close substitutes are available, WTAC and WTP are likely to be closely aligned – as was the case with my favourite brand of breakfast cereal. But perhaps not in the case of my house which, aside from its physical features, also incorporates irreplaceable features such as the friendly next-door neighbours, the carefully manicured garden and the well-tended grave of the faithful family dog.

  Market value as determined by a "willing buyer and willing seller" mental experiment is a convenient benchmark for compensation for regulators and analysts because it is often easily determined from a solid base of transaction evidence. But in the case of unique or highly valued items – my home, family photograph albums, the family dog or life itself – it is likely that a considerable margin will separate WTP and WTAC.

  A decision to build a motorway down the suburban street, demolishing the houses and compensating the owners at "assessed market value" would not accurately reflect the loss to the owners – though those who already have their properties up for sale will be approximately satisfied. For the intra-marginal homeowners, who are not current vendors, the payment necessary to achieve voluntary ("willing vendor") sale would need to exceed the market price. In practice of course the additional amount (Area 1 in Figure 7.1) might be difficult to determine; further, attempts to do so would open the way for strategic behaviour by individual owners to overstate their WTAC and, possibly, concerted action by the owners, or a syndicate of owners, to co-ordinate their negotiating position.

- **Third** there is a reasonable debate about whether a policy change that allows winners to compensate losers *in principle* should also require that the compensation actually takes place. Of course, planning decisions do not have to be adverse. A wider road, a new school, and limits to height restrictions on apartment buildings can have positive impacts on owners of adjacent properties or neighbourhoods. Do the hypothetical, but perhaps very large, benefits to the winners outweigh the costs to potential losers? And even if they do, a public policy challenge is to decide whether the beneficiaries of these windfall gains should or can be charged at least a portion of their windfall gains. Can (and should) the winners actually compensate the losers? An increase in council rates reflecting the rise in property values along a new railway link might cover the costs of the project *and* compensate the losers – a win-win for all parties.

The four measures of compensation

If compensation is to be assessed, the principles that underpin this process are important. The key considerations are

- **The nature of the good or service involved** – is it breakfast cereal or human life? Are there close substitutes – as may be the case with breakfast cereal, less obviously so with a human life or the "right" to healthcare or education?
- **The initial allocation of rights** – where and with whom does the primary right reside – who pays, who receives compensation; is education a "right" or a "privilege"; or health; or sunlight?
- How does "market value" established between willing but not anxious buyers and sellers fit into this analysis?

The foundation for compensation analysis was established by Nobel Prize laureat, Sir John Hicks (1904–1989).[5]

Consider a market regulator seeking to mitigate the impact of a rise in the gas price and compare two policy options.

Policy A: Provide a price subsidy to reduce the cost of gas bills for pensioners, or

Policy B: Provide a direct annual cash payment in *lieu* to pensioners.

Which is the "better" option...better from the pensioner's viewpoint and from the government budget perspective?

Mr Jones is a pensioner. His annual income is OY_1 (Figure 7.2). The price of gas is shown by the slope of his budget line H_1 or the ratio OY_1/OL_3. Mr Jones can choose to purchase no gas, in which case he retains his entire income, OY_1. Or he can spend his entire income on gas, purchasing OL_1. Or he can choose to locate himself anywhere along the H_1 budget line.

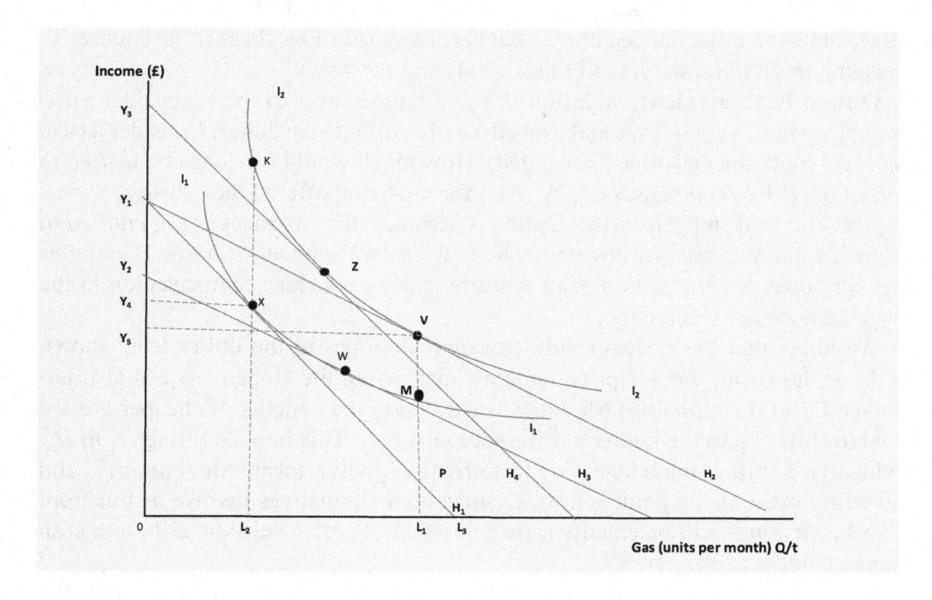

Figure 7.2 The Income/Price Subsidy Trade-Off

Given his income and the price of gas, point X uniquely locates Mr Jones on the highest possible indifference curve, I_1, tangent to H_1. Mr Jones chooses to consume OL_2 units of gas. This gas purchase is an expenditure $OY_1 - OY_4$.

Now consider alternative policies to assist Mr Jones.

Option A, Compensating Variation (CV) – a price subsidy: The reduction in the price of gas (with his cash income unchanged) shifts Mr Jones's budget line to H_2. On this new budget line, the highest achievable utility, I_2 is the point of tangency with H_2, at V. At this new point Mr Jones consumes $L_1 - L_2$ more units of gas than previously and his net cash income (after paying for gas) is also lower by the amount Y_4Y_5. But he is "better off" than previously, as shown by the shift from I_1 to I_2.

Now consider the following thought experiment. How much is this gas price subsidy "worth" to Mr Jones? Or, to reframe the question: How much (in theory) would Mr Jones be WTP in cash for this subsidy?

Given the reduction in the price of gas (shown by the slope of H_2) Mr Jones would be WTP a maximum amount Y_1Y_2 to achieve this subsidy. This payment *at the new subsidised price of gas* shifts the budget line from H_2 to H_3. At this new (hypothetical) income level Mr Jones would choose to locate at *W*. This is the maximum Mr Jones would be WTP because at the new lower gas price he would be as well off (on the original indifference curve, I_1) as he was before.

Implicit in the WTP calculation is the assumption that Mr Jones has no implied "right" to a gas price subsidy. Y_1 Y_2 is however the maximum that he would be WTP to qualify for the price subsidy. Note that through this thought experiment Mr

Jones's income is unchanged at OY_1 but because gas is now cheaper, he chooses to consume more of it (and less of other goods and services).

Option B, Equivalent Variation (EV) – income support: Now consider a different question. Suppose we make an alternative offer to Mr Jones. Consider a cash payment (with the gas price unchanged). How much would Mr Jones be *willing to accept in cash in compensation* (WTAC) for *not* being offered the subsidy?

Note the shift in perspective. Option A assumes that Mr Jones *is not entitled* to a subsidy and we consider how much he is WTP for the benefit. Option B assumes that Mr Jones *is entitled* to a price subsidy and we consider compensation in the event he does *not* receive it.

We know that the price subsidy elevated Mr Jones to the utility level shown by I_2. If, however, the gas price remains unchanged, the slope of H_1, is also unchanged. But if (implicitly) Mr Jones is entitled to the benefits of cheaper gas we need to shift him to the higher indifference curve, I_2. This implies lifting H_1 to H_4. Evidently, a shift in the budget line from H_1 to H_4 will relocate Mr Jones to I_2 and the utility maximising point is now *Z*. Since both alternatives involve a shift from I_1 to I_2, Mr Jones will be equally satisfied with a gas price subsidy, at *V*, or a cash payment equal to Y_3Y_1, at *Z*.

Which is the better policy? Evidently, Mr Jones is indifferent although the quantity of gas consumed has varied as we considered the alternative CV and EV policies. From the market regulator's perspective, however, the differences are significant.

Consider the CV calculation. The price subsidy option is worth Y_1Y_2 to Mr Jones. But the cost to the market regulator is VP – the difference between what Mr Jones would be WTP (Y_1Y_4) and the additional cost to the market regulator at the original gas price; VP > Y_1Y_2.

Evidently, the regulator is paying more than the subsidy is actually worth to the recipient. The gas price subsidy is an expensive way to assist Mr Jones to maintain his existing level of utility or welfare.

Consider the EV calculation. The payment of Mr Jones, Y_3Y_1 is equal to the cost to the market regulator. Note that Y_3Y_1 < VP. Once again, the price subsidy is an expensive option. For a lesser payment, Mr Jones can be lifted to I_2.

In general, price subsidies are more expensive than direct income support payments if the purpose is to assist consumers to achieve a certain level of utility or welfare.

So why are price subsidies so popular?

It is clear that the gas supply company will have a preference for a price subsidy because this will support demand for its product. Price subsidy policies (such as subsidised television licences or bus fares for pensioners) may often be justified by the desire to assist selected groups, but with an additional agenda of influencing their consumption choices. And in the "real world" more complex considerations emerge. What about Mrs Jones? Direct cash payments to families to assist in the purchase of school uniforms or to provide breakfast before school may sound good in theory. But the cash may be diverted to other household consumption priorities less supportive of the children.[6]

How does this work for real estate?

The CV and EV calculations depend on Mr Jones varying his purchase of gas in response to the price of income changes. But some goods and services are not so flexible. Homeownership, for example, is an all-or-nothing situation – you cannot consume more or less of your house. What happens if the quantity to be consumed is fixed? An example might be a twice-weekly meals-on-wheels home delivery service. We continue to employ the Hicks terminology.

Option C – Compensating Surplus (CS) – a price subsidy: Suppose the quantity of gas is fixed.

Then we start from *X*, as before. Now, consider a fall in price. As before, Mr Jones shifts from *X* to *V*. Now pose the question: How much is Mr Jones WTP to secure this fall in the price of gas if we assume that the *new quantity* is fixed at L_1?

This implies a shift from *V* to *M* (not *Y* to *W* as in the CV case). Evidently, Mr Jones is WTP *less* if the quantity is constrained (VM < Y_1Y_2). This makes intuitive sense – a price subsidy is worth less to Mr Jones if his choices are constrained to a predetermined quantity of gas.

Option D – Equivalent Surplus (ES) – an income subsidy: Now as before, pose the question: How much would Mr Jones be *willing to accept in cash in compensation* for not being offered the subsidy *if the quantity of gas purchased is fixed* at the old quantity, L_2? The answer is *XK*. Note that $XK > Y_3Y_1$. Again, this seems intuitively correct. To compensate for foregoing a price fall Mr Jones requires greater compensation if at the same time, he is denied the freedom to adjust the desired quantity of gas consumed.

Three important observations:

First, for a price fall, WTAC < WTP in Figure 7.2 This is a function of the convexity of the indifference curves. In general, for a "normal" good (with a positive income elasticity of demand) the condition WTAC < WTP applies for a price fall. If you are a budget-conscious social architect, it is in general cheaper to support pensioners with a cash payment than a price subsidy. For a price rise, however, the reverse applies. A rise in the price of GP consultations (given our assumptions that the National Health Service is an entitlement for all) may be less costly to the Exchequer that the WTAC that public opinion will demand.

Second, implicit in the discussions of the CV and CS is the proposition that Mr Jones is entitled to *his initial position and gas price*; we then consider, *assuming the price change takes place, how much he is* WTP *to benefit from a lower gas price.* Embedded in the EV and ES discussions is the proposition that Mr Jones is entitled to the *subsequent*, lower gas price; the discussion then centres on appropriate *willingness to accept compensation* assuming the price change *does not* take place. Assumptions about initial entitlement – the current position or a subsequent position lies at the core of many conflicts about payments and compensations. Figure 7.2 illustrates why this is so.

Third, the discussion so far has been confined to the context of a *fall* in the price of gas. The discussion can be repeated in the context of a price rise.[7]

It will be clear by now that market-assessed value based on a hypothetical willing buyer-willing seller thought experiment as a basis for compensation conceals significant complexity from the perspective of both equity (or fairness) and efficiency. The major benefit of assessed market value as the basis for compensation claims in real estate is that it can often be determined with some degree of accuracy based on transaction evidence or analytical calculation. Also evident is that if the object is to achieve a predetermined level of welfare (as proxied by the indifference curves in Figure 7.2) a range of alternative options should be considered.

The all-or-nothing choice

So far this discussion (illustrated in Figure 7.2) has been built around incremental shifts in consumption. A few more hours of work, a few less dollars of travel costs, a modest change in income, an adjustment in the consumption of gas. Often in property markets ownership is not graduated in this way. If your home is expropriated for a new road your ownership falls to zero. In the case of a medical emergency, the alternatives may be life or death. It's an either-or situation.

In this case the indifference curves look like Figure 7.3.

Clearly Figure 7.3 contemplates the possibility where consumption of X is zero. In this case, the Hicks-type analysis still applies,[8] although the shape of the indifference curves can become very steep as they approach the vertical axis. How much are you "willing to pay" – or perhaps "able to pay" is a better descriptor – for

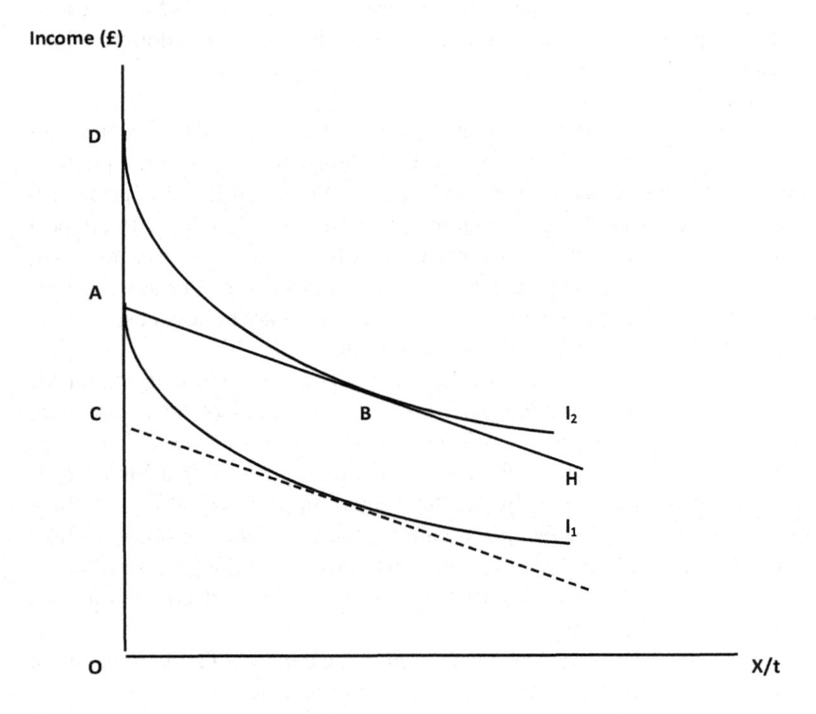

Figure 7.3 The All-Or-Nothing Choice

live-saving surgery? The WTP amount may be limited by your financial resources. How much are you willing to accept in compensation (WTAC) for not having access to the surgery? Do we think the ambulance driver on arrival at a motorway accident should open negotiations with the injured party (a willing and very anxious buyer) for medical treatment? The amount may approach infinity in which case the indifference curves do not intersect the vertical axis at all, contrary to Figure 7.3 – they approach the vertical axis asymptotically.

While WTP and WTAC may differ little for everyday goods – and accordingly market value may be a convenient and pragmatic approximation for compensation purposes – in the case of highly valued goods (and human life is an extreme example) WTAC is likely to exceed WTP by a large amount – this is where debates about pre-existing versus prospective rights become central – the right to clean air, the right to education, the right to medical treatment, compensation for historical mistreatment of ancestors, the claims of cross-border asylum-seekers.

In Figure 7.3 a consumer at point B selects a package of X and income. If X is removed – say the consumer's house is expropriated, utility falls from I_2 to I_1. The consumer would pay an amount of AC to avoid this. Correspondingly a compensation cash payment of AD would return the consumer to his previous level of utility. But suppose we are talking about a human life? In this case, the indifference curves are unlikely to intercept the vertical axis at any feasible level of Y.

Attaching value to a human life is one of the most contentious topics in compensation analysis. Unfortunately, it is also very relevant in many everyday policy decisions. The COVID-19 pandemic has brought this debate into public prominence. While the standpoint that every human life is of infinite value has intuitive appeal, it remains the case that many policy decisions – speed limits on suburban roads, hard shoulders on motorways, capacity limits at hospital emergency facilities, social distancing rules during a pandemic, for example – involve a trade-off, at least implicitly, between dollars and illness, fatalities or injuries.

The Value of a Statistical Life is sometimes employed as a guide to these decisions.

> Suppose each person in a sample of 100,000 people were asked how much he or she would be willing to pay for a reduction in their individual risk of dying of 1 in 100,000, or 0.001%, over the next year. Since this reduction in risk would mean that we would expect one fewer death among the sample of 100,000 people over the next year on average, this is sometimes described as "one statistical life saved." Now suppose that the average response to this hypothetical question was £100. Then the total dollar amount that the group would be willing to pay to save one statistical life in a year would be £100 per person × 100,000 people, or £10 million. This is what is meant by the "value of a statistical life."[9]

Empirical research suggests that public policy regulators implicitly apply widely different values to a "statistical life" depending on the context – road engineers designing the shoulder on a busy motorway may apply very different metrics from epidemiologists advising governments during a pandemic. For example, in 1993 aircraft cabin protection against fire cost US$200,000 per life saved. Automobile side door protection cost US$1.3 million per life saved and US Environmental Protection Agency (EPA) asbestos regulations saved lives at a cost of US$104.2 million each.[10] Nor are these life-or-death choices confined to public policies. Every automobile designer or ambulance driver knows that sometimes a choice has to be made between safety and service.

Rent-seeking – the other side of the compensation coin.

Policymakers and regulators can impose costs on individuals and societies; they can also distribute benefits. Public house trading licences, taxi licences, zoning rights and building approvals can confer huge benefits to selected individuals. One transparent way to value these rights – and to allocate them – is by public tender or auction. The winner, WTP the highest price, one might assume is also the applicant best able to extract maximum value from the project or service. There are potential efficiency gains from contracting out public services such as garbage collection, school lunches, postal services or gas meter reading to the highest bidder. But there are risks as well.

The Roman Empire, for example, auctioned off the right to collect taxes in certain jurisdictions, most famously in Judea in the year 6 BC. The winning bidder from these auctions then had a financial interest in extracting the maximum revenue from that community – one reason why tax collectors were so unpopular and why they get such a bad rap in the New Testament:

> Now it happened, as He was dining in Levi's house, that many tax collectors and sinners also sat together with Jesus and His disciples; for there were many, and they followed Him. And when the scribes and Pharisees saw Him eating with the tax collectors and sinners, they said to His disciples, "How is it that He eats and drinks with tax collectors and sinners?"[11]

Public and transparent auctioning of taxation licences across the Roman Empire had much to commend on efficiency grounds. Often, however, successful tenders for the provision of public services require additional attributes beyond the financial capacity of the highest cash bidder. This might be, for example, demonstrated track record or financial capacity, gender or racial profile, and environmental credentials.

Less admirable may be financial contributions to a political party or an expenses-paid holiday in Majorca for the chair of the local Council amenities committee.

Since these benefits are available and can be very large, we can expect competing parties to engage in strategic behaviour to access these advantages. Some of these activities may be legal, even desirable. Other activities may unethical or even illegal. In either case, efforts to access these benefits ("economic rents") by transferring wealth rather than creating it is termed "rent-seeking." Rent-seeking absorbs resources with some expectation, but no guarantee, of a favourable outcome. It's a gamble. The Majorca holiday may not result in the award of a trading licence.

And we have reason to wonder about Levi's motives in extending his hospitality to representatives of the local Inland Revenue Service as well as Jesus, as described by the apostle Mark. Was his prospective payoff in this world – or the next?

Consider the following thought experiment.

You have been selected as one of 50 delegates to attend a two-day in-house sales training program. At the commencement of proceedings, the Director of Sales brandishes an envelope containing £1,000.00 in cash. This will be awarded to the highest bidder during the final session on the afternoon of Day Two. Participation in the auction is voluntary. If you wish to participate, place your own bid (in cash) into an envelope and hand it to the Sales Director sometime before luncheon on Day 2. At the final session, the Sales Director will open the envelopes he has received, awarding the prize of £1,000.00 to the highest bidder.

But this auction comes with a catch – the bidders will not receive their cash back.

Try this at your next sales conference.

What do you expect to happen?

Presumably, no one will bid more than £1,000.00 – although in a room full of ultra-competitive aspirant salespeople, each trying to catch the Sales Director's eye, anything is possible. At the other extreme all 50 attendees might agree during the cocktail session on the evening of Day 1 of the conference to form a consortium. They will submit a single bid of £1 and split the proceeds, each receiving a net amount of £19.98. But this consortium will be unstable. Jane Smith might secretly decide to break the consortium agreement, submit a personal bid for £2 and scoop the pool. But if Jane Smith can figure out this, so can Fred Brown. And he might bid £3 to be more certain. A secret bid of £900 from Joe Green would be fairly certain of winning, with boasting rights and a £100 payoff – but at some risk. After all, Anne Wilson, anticipating Joe Green's action, might submit a bid of £901.

This is a bidding game without a solution – the Sales Director may receive one bid for £1 if the consortium holds – but if several players choose to bid £900 the Sales Director may receive substantially more than the £1,000.00 on offer; he might earn a large profit. (To avoid debate, repeated real-world experience of this game suggests to this author that the former is the more likely outcome at corporate and academic events; but this may be a reflection of the mutually – supportive environment where these experiments have been conducted.)

In this thought experiment the Sales Director is in the position of a regulator handing out a concession to, for example, provide a monopoly bus shuttle service between the airport and the city centre; or a cleaning contract for government offices. Competitors seeking this concession will engage in activities to win the concession. These activities may be perfectly legal and productivity increasing – for example, investing in fuel-efficient shuttle buses or modern cleaning equipment. Other activities may be designed to meet other criteria – for example, hiring staff to meet the city government's racial or gender quota targets. A donation to the mayor's re-election fund might be effective. This will do nothing to enhance productivity – it is a straight financial transfer better described as a bribe. In all cases, however, there is no guarantee of a return on the investment – the same calculation made by Jane Smith, Fred Brown and Joe Green. As with the sales conference example, the financial resources committed by all competitors to winning the shuttle bus tender may be less, or significantly more, than the value of the concession itself. Clearly the position of Sales Director, with its agreeable perquisites, will be sought after and resources are likely to be allocated to achieving this position.

The literature and measurement of rent-seeking have been extensive since Gordon Tulloch (1922–2014) defined rent-seeking in 1967.[12] As might be expected, the evidence is that in some situations substantial resources are devoted to economically wasteful rent-seeking behaviour. At the microeconomic level, the proliferation of political lobbyists, for example, suggests a wasteful allocation of resources. At the macroeconomic level Mancur Olsen, for example, has described the adverse impact of rent-seeking, and related activities, on whole societies.[13] In feudal Medieval Europe there was a trade-off: cultivate the land or cultivate the lord of the manor. The result was a low-growth, largely stagnant society.

Barriers to change – changing bad laws is hard, but possible

A new law, or any change in existing law or regulation, confers costs and benefits on at least some people. Laws that outlaw murder impose costs on prospective murderers. The losers and potential losers from any proposal to introduce a new or amended law can be expected to oppose the change. The prospective winners are likely to support the change. But the deck is not always evenly stacked symmetrically between winners and losers. The potential losers from a proposed change in laws or regulations often have concrete evidence of their likely loss. In contrast, the potential winners have only prospective and as-yet unrealised gains. A high-speed rail line through prime farmland may be opposed by the local communities who have solid evidence of their anticipated financial and immediate psychic loss. The larger number of future commuters who will benefit from time and cost savings have only hypothetical gains to contemplate and are probably less well organised. That is one reason why regulatory change is fraught with difficulty, and progress is uncertain.

Import tariffs are a clear case of winners and losers. A tariff on imported cars raises the price of cars to motorists in the importing country, while protecting the profits of domestic car manufacturers, and the jobs of workers in that industry.[14] Car manufacturers are likely to be well-funded and organised. A unionised group of car workers will be similarly coordinated, often with financial resources. Current and future motorists, and their passengers, probably outnumber the workers who benefit from of import tariffs, but individually each motorist has less financial incentive to organise opposition to import tariffs, and the costs of doing so are likely to be high.

Once import tariffs are imposed, the removal of tariffs is even more costly from a political standpoint. A worker about to lose a lucrative job in a car plant has a strong incentive to actively oppose the proposed tariff reduction. A motorist, who looks forward to a reduction in the price of her next car, partly offset by a reduction in the trade-in value of her current vehicle, is less likely to join a protest march to Parliament advocating tariff reduction.

Nor is it obvious that the state will be a supporter of change or even act as an unbiased umpire. Often the government itself benefits from the revenue arising from the licence fees and tariffs that limit competition. It is not only in corrupt regimes that government officials and politicians are tempted with financial incentives and other perquisites of office to introduce or maintain taxes and regulations that impose costs on the broader society.

Consider the following (real life) advertisement:

> ***MELBOURNE: For Sale – Taxi Plates***
> *Date Listed: 2013-07-13*
> ***Taxi licence for sale***
> *Unrestricted Melbourne metro taxi licence for sale for AU$350,000. Secure investment for the future with the current lease at AU$3,000 until 2015. Call on xxxxxxxxx*

In the Australian city of Melbourne, the number of taxi licence plates is limited to 5,090 (as of May 2011). The value of a taxi licence plate to operate in Melbourne, the capital of the state of Victoria, had more than doubled over the previous decade.[15] Many vested interests are committed to preserving the regulatory *status quo*, including:

- The City of Melbourne, which benefits from the fees for the sale of licences;
- Cabcharge, a national company that earns fees from the compulsory EFTPOS taxi credit card payments system;
- Manufacturers and suppliers of mandated taxi equipment such as meters and cameras;
- Taxi licence brokers who trade the licence plates;
- Existing owners of taxi licence plates.

The Taxi Industry Inquiry (2012) estimated that Melbourne taxi users (customers) as a group pay around $120 million per year to maintain the value of taxi licence plates.

Do they get value for money?

> …Victoria's taxi industry has operated for many years as a 'closed shop', with a small number of licence holders protected from the effects of competition at the direct expense of consumers, taxi operators and taxi drivers (who continue to experience low levels of remuneration, poor working conditions and a highly risky work environment). The inquiry found no public interest or other grounds for allowing this situation to continue.[16]

Among the costs of the system, driver turnover is high, requiring the training of 1,500 to 2,000 new drivers to be accredited each year. This is costly, even though many visitors to Melbourne will attest that the standards required for accreditation as a cab driver do not appear to be especially arduous.

Outright deregulation of the Melbourne taxi licence system would lead to a more efficient allocation of resources and a better service for travellers. Deregulation of the fare structure would allow taxis to offer a wider range of services tailored to customer preferences, including safer vehicles and more experienced drivers, as well as a basic, and therefore inexpensive, service for people simply wanting to get from one place to another.

Compensation: How to lubricate the wheels of change

However, total deregulation would impose an immediate capital loss of around AU$350,000 on each owner of a taxi licence plate. We can therefore anticipate that owners of licence plates would vigorously oppose any proposal to deregulate the industry.

And the owners have some moral rights on their side. Having invested a considerable sum in this asset (financed perhaps by a mortgage on their home) on the basis of an implicit or explicit assurance by the City of Melbourne that the scarcity value of taxis would be maintained, deregulation would represent a significant change to the rules of the game.

When the "rules of the game" are changed there is a case for compensation even if the rules themselves impose net costs on society. The case for compensation rests on the moral argument that substantial investments may have been made, time committed and sacrifices incurred to acquire tangible assets (such as physical property or a motor vehicle) or intangible assets (such as the right to operate a taxi or to sell liquor or to practice medicine) on the basis of an explicit or implicit undertaking that these rules would be permanent.

A more pragmatic case for compensation is that opposition to the proposed change is likely to diminish if reasonable compensation is on offer.

And, importantly, it can be shown that if the proposed change is, in fact, in the direction of greater efficiency, the winners from the change can, in principle compensate the losers and still come out ahead – it's a win-win situation for all parties or a *pareto-superior* outcome.

This is illustrated below in the illustrative discussion below:

According to the Government of Victoria Taxi Services Commission report (2012) there were 5,090 taxi plates in the City of Melbourne. Let's assume a market value of AU$350,000 per plate. This represents a capital value of AU$350,000 × 5,090 = AU$1,781.5 million. Since the market price set by the latest transaction is AU$350,000, this is KN in Figure 7.4. The Area 2 + 3 = AU$1,781.5 million. Note that the capital value of all licence plates does not capture the entire profits of the taxi operator. An upward sloping Supply line implies that some operators have lower costs than others due to greater efficiency or other advantages, such as proximity to the central business area. The entire profits for the 5,090 licence plate

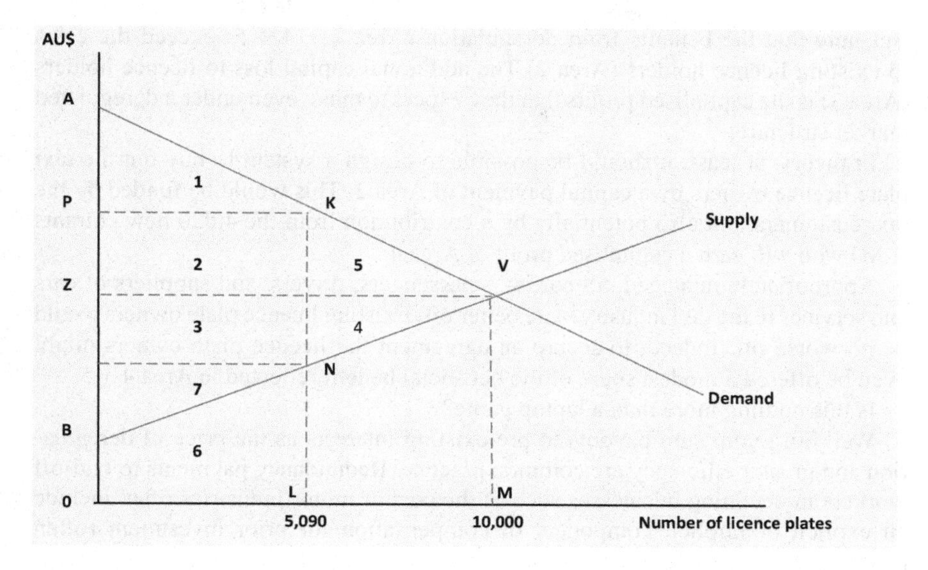

Figure 7.4 Measuring the Impact of Taxi Licence Plate Regulation

holders are represented by Area 2 + 3 +7. The area under the supply line, 6, represents operating costs. Note that the upward slope of the supply line implies that operating costs vary across the industry. In the absence of the taxi licencing system, Figure 7.4 implies 10,000 taxis on the roads of Melbourne. The Taxi Commission report estimated an annual cost to consumers of AU$120 million to maintain the value of these licence plates. If we capitalise this annual cost at, say, 5 percent, this implies a capital cost to consumers of AU$2,400 million, represented by Area 2 + 5. Area 2 is therefore AU$1619.09 million; Area 5 is AU$790.91 million.

Deregulation of the taxi industry implies a reduction in the price (or cost) of travelling by taxi. This might emerge as lower fares or as improved service, shorter queues for taxis in wet weather, reduced waiting times or a more courteous driver. Existing plate owners (the current 5,090 licence holders) would suffer a capital loss equivalent to Area 2, a straight transfer to taxi customers. However, these existing operators would continue to earn a profit, shown by Area 3 + 7. This may not be transferable as a capital sum, however. Area 3 + 7 might reflect the skill and know-how of the existing operators, their relationship with regular passengers (potentially an important asset for Melbourne taxi drivers) or simply hard work.

Deregulation allows customers to benefit from lower fares (a fall of PZ), which also captures a potential improvement in the quality of taxi services. In addition, customers would gain Area 5 as additional taxis entered the industry. These new taxi operator entrants would also earn profits, Area 4, up to the marginal, 10,000th entrant, who will break even with a capitalised annual income of OZ.

Figure 7.4 suggests a straight transfer from existing taxi operators to customers, as well as benefits to potential additional customers and new taxi operator entrants.

But note that the benefits from deregulation (Area 2 + 4 + 5) exceed the costs to existing licence holders (Area 2) The additional capital loss to licence holders (Area 3) is the capitalised profits that they expect to make even under a deregulated market structure.

In theory, at least, it should be possible to design a system to buy out the taxi plate licence owners by a capital payment of Area 2. This would be funded by the taxi customers, but also potentially by a contribution from the 4,020 new entrants (LM) who will earn a capitalised profit of Area 4.

Appropriately managed, all parties – passengers, drivers, and suppliers of cars and services to the taxi industry – are better off. Existing licence plate owners would be no worse off. Indeed, to secure an agreement the licence plate owners might even be offered a modest share of the net social benefit reflected in Area 4 + 5.

Is this nothing more than a laptop game?

Well no. Lump sum payouts to pre-existing interests as the price of deregulation and greater efficiency are common practice. Redundancy payments to laid-off workers in declining businesses such as the coal or motor industries often include an explicit or implicit component of compensation for prior investment (often proxied by time served in the industry).

In 1834 the United Kingdom Slavery Abolition Act took effect, abolishing slavery in most British colonies throughout the British Empire. Slave owners were offered compensation amounting to £20 million pounds, around 5 percent of the UK GDP at that time. To finance this measure the UK government raised a loan (that was technically not finally repaid until 2015). While the ethics of payments to slave owners – some already wealthy – and the method of distributing these monies (which required personal attendance in London) may be debated, the effect of the measure was to limit objections and obstruction by politically influential slave owners throughout the Empire. The positive economic payoff (moral issues aside) from the abolition of slavery was a more mobile, motivated and productive labour force. Harder to quantify financially was personal liberty for tens of thousands of people through the British Empire. Failure to adopt this expensive compensation measure would certainly have raised stronger objections from slave owners who would have been in the same position as Melbourne taxi licence plate holders. The compensation measure was indeed costly; but failure to abolish slavery in British colonies in early 1834 might have been far more expensive, as the American Civil War (1861–1865) perhaps bears witness.

The essence of this analysis is to demonstrate that inefficient regulations and laws can be circumvented in ways that do not necessarily set losers in an antagonistic position to winners – reform and deregulation do not have to be a zero or negative sum game with the conflict and delay that this often entails. A win-win outcome can in principle be achieved for all parties but only if the proposed shift is towards greater economic efficiency. It is this shift that creates the financial resources required to compensate the "losers" while retaining benefits for the "winners."

Appendix 7A – The four measures of compensation

Table 7.1A sets out the full scenario for price rises and falls. The shifts refer to Figure 7.2.

Table 7.1A Compensation paid and received – the four scenarios

CV – entitled to pre-change situation	*Change takes place*	*Shift*
X variable		
Therefore		
	Price rise – WTAC for tolerating price rise	V ⇒ X ⇒ Z
	Price fall – WTP rather than forego benefits of price fall	X ⇒ V ⇒ W
EV – entitled to post change situation	Change does NOT take place	Shift
X variable		
Therefore		
	Price rise – WTP to avoid price rise	V ⇒ W
	Price fall – WTAC to forego benefits of price fall	X ⇒ Z
CS – entitled to pre-change situation	Change takes place	Shift
X fixed at NEW quantity		
Therefore		
	Price rise – WTAC for tolerating price rise	V ⇒ X ⇒ K
	Price fall – WTP to secure a price fall	X ⇒ V ⇒ M
ES – entitled to post change situation	Change does NOT take place	Shift
X fixed at OLD quantity		
Therefore		
	Price rise – WTP to avoid price rise	V ⇒ M
	Price fall – WTAC to forego benefits of price fall	X ⇒ K

Notes

1 The film *The Castle* (1997) offers a colourful but informative perspective on the process and politics of compulsory acquisition of residential properties adjacent to an airport.
2 The important distinction between *quantity demand* and *quantity demand per time period* is explored in Chapter 9.
3 The key word here is *unexpectedly*. If the consumer anticipated the arrival of the additional quantity the optimal strategy would be to delay the purchase of *OP*1 and then to pay *OP*2 for *OQ*1 + *Q*2 − *Q*1.
4 This is also known as *first-degree price discrimination*. This is difficult to achieve in practice, but retailers are adept at "volume discounts" or second-degree price discrimination – one bottle of wine, £10, a case of six bottles, £50. Effectively you pay £10 for the first bottle and £6.66 each for the other five.
5 Hicks, J. R., The Four Consumer's Surpluses, *Review of Economic Studies*, 1 (1943), pp. 31–41. See also an extensive analysis of cost-benefit analysis techniques and applications in OECD, 2000. Willingness to Pay vs. Willingness to Accept, in *Cost-Benefit Analysis and the Environment: Recent Developments*, OECD Publishing.

6 This analysis (which is a standard textbook treatment) demonstrates the superiority of income compared to price subsidies. It provides a useful starting point and valuable insights into a discussion on subsidies for selected groups of people. But it is a partial analysis. Milton Friedman 1967 (*Price Theory: A Provisional Text*, Chicago Press, Chapter 2) presents a more comprehensive analysis.
7 Appendix 7A defines the four compensation measures for price rises and falls.
8 Technically the indifference curve convexity assumption is violated by the intersection with the vertical axis.
9 United States Environmental Protection Agency. See also *Best Practice Regulation Guidance Note: Value of Statistical Life*, Department of the Prime Minister and Cabinet, Australian Government. October 2018. https://pmc.gov.au/sites/default/files/publications/value-of-statistical-life-guidance-note.pdf.
See also an illustrative analysis: *Differences between Willingness to Pay and Willingness to Accept for Visits by a Family Physician: A Contingent Valuation Study*, BMC Public Health, 2010. https://bmcpublichealth.biomedcentral.com/articles/10.1186/1471-2458-10-236.
10 Viscusi, W. K., The Value of Risks to Life and Health, *Journal of Economic Literature*, 31, no. 4 (1993), pp. 1912–1946.
11 *The Bible* (The New King James Version), Mark 2: 15–16.
12 Tullock, G, The Welfare Costs of Tariffs, Monopolies, and Theft, *Western Economic Journal*, 5, no. 3 (1967), pp. 224–232. See also Tullock, G., Efficient rent-seeking, in Buchanan, J., Tollison, R., and Tulloch, G. (eds.) *Toward a Theory of the Rent-Seeking Society* (College Station, TX: A&M Press, 1980), pp. 97–112.
13 Olsen, M., *The Rise and Decline of Nation: Economic Growth, Stagflation and Social Rigidities* (New Haven: Yale University Press, 1982).
14 An import tariff that is a fixed amount, say £1,000 per car or £10 per bottle of wine, encourages the import of luxury vehicles and expensive imported wine – the so-called Third Law of Demand because the more costly the item the lower is the effective tariff rate.
15 *Customers First – Service, Safety, Choice*, December 2012, Taxi Services Commission, ISBN 0-7311-8796-2, p. 4.
16 Taxi Services Commission, 2012, p. 4.

Bibliography

Department of the Prime Minister and Cabinet, Australian Government, October 2018. *Best Practice Regulation Guidance Note: Value of Statistical Life,* s.l.: s.n.

Friedman, M. 1967. *Price Theory: A Provisional Text.* Chicago, IL: Chicago Press.

Hicks, R. J. 1943. The Four Consumer's Surpluses. *Review of Economic Studies,* 1, pp. 31–41.

Martín-Fernández, J., del Cura-González, M. I., and Gómez-Gascón, T. 2010. Differences Between Willingness to Pay and Willingness to Accept for Visits by a Family Physician: A Contingent Valuation Study. *BMC Public Health,* 10, p. 236.

OECD. 2006. Willingness to Pay vs. Willingness to Accept, in Pearce, D. Atkinson, G and Mourato, S. (Eds.) *Cost-Benefit Analysis and the Environment: Recent Developments.* s.l.:OECD Publishing.

Olsen, M. 1982. *The Rise and Decline of Nation: Economic Growth, Stagflation and Social Rigidities.* New Haven, CT: Yale University Press.

Ryan, W. J. L., and Pearce, D. W. 1977. *Price Theory.* London: Macmillan.

Taxi Services Commission. 2012. *Customers First – Service, Safety, Choice. Draft Report,* Melbourne, Vic: Taxi Industry Inquiry.

The Bible (The New King James Version), Mark 2: 15-16, n.d.

The Castle. 1997. [Film] Directed by Rob Sitch. Melbourne, Australia: Working Dog Productions.

Tullock, G. 1967. The Welfare Costs of Tariffs, Monopolies, and Theft. *Western Economic Journal,* 5(3), pp. 224–232.

Tullock, G. 1980. Efficient rent-seeking, in Buchanan, J., Tollison, R. & Tullock, G. (Ed.) *Toward a Theory of the Rent-Seeking Society,* A&M Press, College Station, TX, pp. 97–112.

Viscusi, W. K. 1993. The Value of Risks to Life and Health. *Journal of Economic Literature,* 31(4), pp. 1912–1946.

8 When property rights fail – externalities

Good fences make good neighbours

Property ownership and occupancy confer a package of rights and obligations. These rights, however explicit or extensive, are always subject to limits. Some rights, such as the boundary of the land upon which your home is located, may be precisely defined. Other rights, such as the right to privacy, sunlight or clean air, are often less well defined and can give rise to *externalities*. Externalities arise when the actions of one party impose costs (or benefits) on a second party that are not directly captured in the price system. Externalities may be positive or negative. They may add value or subtract it. A factory pumping polluted water into a river that kills local fish stocks is an obvious example of a negative externality. These external costs or benefits are not (at least in theory) taken into account by the party responsible (but, as we shall see, the real world offers counterexamples).

Often dismissed as rare or exceptional, externalities are in fact pervasive. They are central to debates about climate change and carbon dioxide emissions, for example. Unsurprisingly it is negative externalities (such as noise from a nearby airport or smoke from a steel mill) that attract the greatest attention. Negative externalities are often interpreted as a case for legal action or compensation to be paid by the "offending" to the "innocent" party. A better perspective is that the costs and benefits of externalities are reciprocal. Therefore, the fundamental understanding of the causes and logically coherent solutions to externality problems is to be sought in the realm of economic analysis. Progress is to be sought neither in debates about moral positions nor in a search for legal precedents.

Economic analysis offers a range of alternative insights and suggested strategies to arrive at *efficient* and, more arguably, *equitable* outcomes where externalities occur. Market-based solutions to externality problems can sometimes be an efficient alternative to regulatory or legal interventions that can involve costly delay, fees, taxes or the imposition of penalties and compensation payments.

DOI: 10.1201/9781003111931-8

Introduction: Externalities...externalities...wherever you look

Suppose you look upwards. Standing on your front lawn, you watch an aircraft high overhead. The aircraft is unhindered, but a drone circling at a much lower level above your house might give you cause for legal action. How far upward from your lawn does your property ownership extend; and how far downward into the Earth? The local council can decide to expand the local school, narrow the road to install a bicycle path, limit local parking or reduce neighbourhood police surveillance. The fire brigade, the gas meter reader, or police armed with a search warrant all have access to your home under defined circumstances. All these actions, none of which you directly control, can have an impact on you, the market value and your enjoyment of your property.[1]

Technically, "externalities" arise when private and social benefits or costs diverge. Often, externalities are treated exclusively in a negative context – the smoke from a factory, the noise near an airport. But externalities can be positive, too. The smoke and the aroma from your neighbour's backyard barbeque might be an example. Both smoke and aroma drift across the fence into your garden. You may enjoy the aroma but dislike the smoke. You can't have one without the other, but you will experience both. Your neighbour's Bank Holiday culinary planning probably does not take account of your preferences and he may not even have the courtesy to extend you an invitation to attend. This could be a mistake. An invitation to join the party might be a way for your neighbour to "internalise" the externality that the barbecue potentially creates for the adjacent properties.

Early economists and some standard economic texts treat externalities as exceptional (and perhaps rare) events arising from the imprecise specification of formal property rights. These problems can then be addressed by one-off actions such as better legal drafting, regulations or taxes imposed on the "offending" party, once the source of the "nuisance" has been identified.

In fact, externalities are pervasive, as a stroll down your local High Street will reveal – the coffee aroma, the offensive words on a passing T-shirt, the noise from a construction site, the pavement busker... these impressions are all freely available, indeed, compulsory. The sources of some of these experiences – the barista and the property developer for example – take no account of you and your preferences, and it's hard to see how they could, even if they wanted to. Others – the student with the T-shirt and the pavement musician – do perhaps have you in mind. Maybe the barista does too – the smell of freshly ground coffee is a strong marketing signal. And perhaps the presence of the busker, hovering near the entrance to the Khardomah, is no coincidence either.

Societies differ widely in how they identify externalities and how they manage them. Although externalities occur everywhere, from palaces to prisons, externalities arising in a real estate context are heavily represented in the law courts, in backyard disputes and in local government forums. We will consider first the technical analysis of externalities. Then we discuss the two dominant schools of thought on policies to manage externalities – associated with A C Pigou[2] and R H Coase.[3] Finally, we consider some real-world analysis and implications in real estate markets.

Negative externality – the textbook case

Figure 8.1 illustrates the classic textbook case of a negative *externality*. A cotton mill generates smoke as a by-product of its fabric manufacturing activities. The smoke from the mill imposes physical, financial and perhaps psychic costs on the residents in the adjacent housing estate.

Figure 8.1 defined the financial parameters of production for the cotton mill. The horizontal axis measures the Quantity Produced per Time Period or the *Rate of Production, Q/t.*[4] PP' is a Price line; it traces a set of market prices (*P*) against a set of production rates (*Q/t*). Therefore, if the production rate is Q_1 per month and the market price is P_1, the total monthly revenue received by the enterprise is OP_1BQ_1.[5]

The mill's Marginal Cost is shown by line PMC. The marginal cost is defined by current technology – this is, PMC traces how costs rise with the rate of cotton production. This is the *private marginal cost* incurred directly by the enterprise in its production activities. The orthodox economic analysis argues that in selecting the profit-maximising rate of production the enterprise will take account only of its private costs. The output of the mill sells at the prevailing market price OP_1.

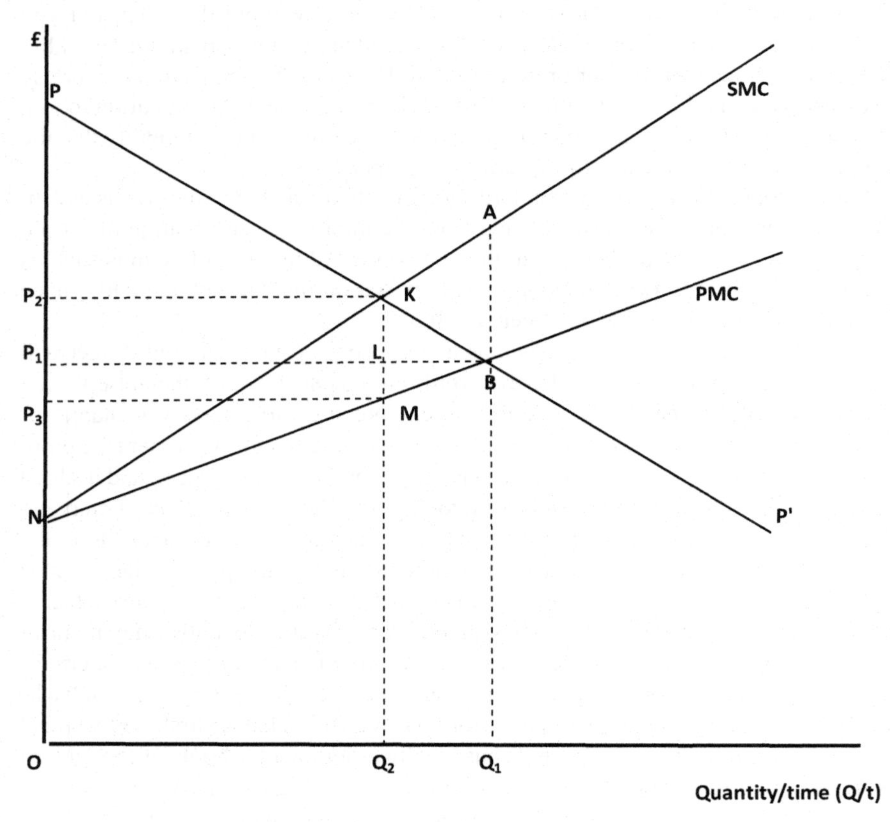

Figure 8.1 The Principle of Externalities – The Negative Case

The mill has no control over the market price. But given this price, the profit-maximising rate of production for the enterprise, given by the intersection of PP' and PMC, will be OQ_1.

The Total Cost incurred by the enterprise is the area under PMC.[6] At production rate Q_1 the total production cost per time period is the area under the PMC line, $ONBQ_1$. Total Revenue for the enterprise is identical to Total Expenditure by the consumers.

At the production rate Q_1 the enterprise profit is

Total Revenue *minus* Total Cost
OP_1BQ_1 *minus* $ONBQ_1 = NP_1B$

However, PMC captures only the *private costs* incurred by the enterprise. In addition, there are *social costs* or *externalities* that arise (we assume here) from the smoke, inextricably linked to cotton production. The difference between PMC and the social marginal cost line, SMC, shows the *additional* costs imposed by the mill on the residents of the adjacent housing estate. For each additional square meter of cotton fabric produced beyond OQ_2, the cost of production including social costs, SMC, exceeds the private marginal cost, PMC, of the mill.

From a broader social perspective, taking into account the additional external costs arising from the operations of this enterprise, the optimal rate of production is OQ_2. If production exceeds OQ_2 the cost of each additional unit (private *plus* social cost) shown by SMC exceeds the market value of that unit shown by PP'. The additional net social cost arising from production OQ_1 instead of OQ_2 (the amount by which the additional total cost exceeds the value created) is shown as area KAB.

Four points are important:

- **First**: in Figure 8.1 we have considered two alternative rates of production, OQ_1 and OQ_2. However, a reduction in the rate of production is not an isolated event. The market price, enterprise profits and consumers themselves are all potentially affected by a change in production rate. How are the costs and benefits of this lower rate of production apportioned? In the real world, the answer to this question is central to policy debates. We examine this important question in the two Sections below, but for now, consider the impact of a simple regulation limiting production by this mill (and *only* this mill) to a rate of OQ_2. Since we have assumed this mill to be a price taker in the cotton fabric market the market price remains unchanged at OP_1. The enterprise profit is now NP_1LM, a decline of LBM. Evidently, the enterprise is now less profitable. Householders in the adjacent area, however, are better off. Since the market price of the fabric is unchanged (OP_1), consumer surplus is also unchanged (P_1CB). They purchase an additional Q_2Q_1 fabric from other cotton mills not subject to the production limit imposed on this mill. The real-world application? In a competitive market, enterprises may lobby to impose "standards" that have the effect of limiting, even eliminating, competitors. Safety standards, environmental standards, and hygiene standards are all barriers to entry advocated by technologically advanced enterprises to the disadvantage of their competitors – and imposed, of course, with the best of intentions!

- **Consider an alternative scenario**: Suppose that this enterprise is representative of *all* cotton mills and a regulation is introduced to limit production of each mill to OQ_2. Under this plan, industry output will fall and for a given level of demand the market price of fabric will rise to, let's say, OP_2. Now the payoffs are rather different. Consumer surplus – the net benefit to customers of the mill industry declines from P_1CB to P_2CK. If the reduction in consumer surplus, $P_1P_2KB < KAB$ customers of the mills are net worse off – although we must be cautious that the customers of the mills are not identical with the householders adjacent to the mills. There are distributional effects. Further, enterprise profits for the mill industry (this being a representative cotton mill) *increase* by P_1P_2KL minus LBM. Regulation to limit production can be an effective transfer from consumers to producers. It is not always the business community that is opposed to regulation – indeed the cotton mill owner may be an enthusiastic advocate of measures to protect the environment by limiting production for good commercial reasons. It is not surprising therefore to find that commercial companies are sometimes at the forefront of policies to regulate markets, including environmental regulation. Owners of tall city-centre office buildings may favour height limits on future developments; shopping mall owners may favour the imposition of zoning regulation to preserve local amenities, limit traffic congestions and, indirectly, protect existing retailer tenants from future competitors. It may be difficult to disentangle the so-called "social licence" from financial self-interest. As a qualification, it should be noted that the increase in profit to the cotton mill from a limit to production is probable, not inevitable. It is the outcome of the way Figure 8.1 was constructed. Assessing the burdens and benefits of externalities, as well as measures to address them, is often more complex than appears at first glance. Claims to represent, or to benefit, the broader community are not necessarily in conflict with financial self-interest, though they may conflict with the interests of the broader community.
- **Second**: While OQ_2 is shown as the socially optimal rate of production in contrast to the unregulated or "laissez-faire" outcome of OQ_1, this outcome reduces but does not eliminate social costs. External costs of NKM are still incurred even at the "optimal" production rate OQ_2. *This is important.* We are not required, nor are we able, nor indeed would we wish, to live in a world entirely without noise, smoke and pollution. Market prices and costs (taking into account private and social dimensions) establish an *optimal*, not a zero, burden of externalities. Driving smoke pollution to zero would also drive the rate of production (Q/t) to zero. Advocates of zero pollution or carbon emission policies do well to take note of the costs as well as the prospective benefits of the policies they propose.
- **Third**: Note, and this is not a trivial observation, that the customers of the enterprise may not be identical with the immediate neighbours who suffer the impact of the smoke. Your local cycling club does not live in the cloud of smog that hangs over Guangzhou where their bicycles and high-viz thermal outfits are perhaps manufactured. Neither the owner of the enterprise nor the customers of

the cotton mill (if they are not also neighbours) are interested in the additional social costs arising from the smoke. However, in an efficient world, taking account of both private and social costs and ignoring second-order distribution impacts, the rate of production would be OQ_2.

- **Fourth***:* We were careful to define *PP′* as a Price line, not a Demand line, and to assume that the cotton mill is a price taker in the cotton fabric market. Had we defined *PP′* as the Demand line for the cotton mill, we would be in the world of monopoly analysis. The cotton mill would maximise profit where Marginal Cost (MC) = Marginal Revenue (MR). The MR line is not shown, but we know that the rate of production would be less than OQ_1. It might be less than OQ_2... or more than OQ_2; it might even be identical to OQ_2. In a world of production externalities, monopoly enterprises by reducing production (and negative externalities) can make a positive contribution towards achieving a socially optimal outcome – an uncomfortable result for advocates of competitive markets.

Of course, the principles of externalities are not limited to economic "goods." Economic "bads" operate according to the same rules. Illegal drug use, for example, imposes long-term costs on the users and on society at large. It is at least worth considering whether you would prefer the illicit drug business in your neighbourhood to operate on a competitive basis or be managed by a local drug lord who would limit supply and reduce consumption by driving up the price. The evidence of turf wars between competing drug syndicates in some cities suggests that the monopoly profits arising from territorial control in urban areas can be significant. This implies that localised illegal-drug monopolies are indeed generating a positive externality by limiting drug availability via the price system – although this does not preclude other adverse personal and social impacts from their illegal activities – an example of negative consumption externalities.

Perspectives on externalities, and how to manage them, fall into two broad schools of thought, associated with A.C. Pigou and R. H. Coase.

The Pigou approach – call in the regulators

A.C. Pigou (1877–1959) saw externalities through the lens of *market failure*. In his analysis, prevailing property rights sometimes fail to take into account the "spill-over" effects arising from production, transport or other activities. The corrective to this was government intervention in the form of taxes or subsidies, although Pigou did not prescribe detailed policies to address specific problems. Implicit in the Pigouvian perspective is the notion that externalities are isolated and rare, to be treated on a case-by-case basis.

In the section above we considered a simple regulation limiting the rate of production. But a range of other policy options are available. For example, a tax on production on a sliding scale rising from zero to a rate of AB per unit at production rate OQ_2 would achieve the required result. In the case of a progressive production tax the firm's profit changes from NP_1B to NP_2K (which may or may not constitute an increase) and the revenue raised by the tax would be NKM. Note that although the socially optimal rate of production (OQ_2) is achieved the burden (or incidence) of the tax would be spread between the customers (who lose P_1P_2KB) and the mill owner who may, or may not, earn lower profits. An important principle here (one that is often overlooked in public debates on taxation and regulatory matters) is that although the enterprise pays the tax, the *incidence of the tax* may be elsewhere – the hand that signs the cheque may not be attached to the entity that bears the burden of the tax.

In general, the Pigouvian prescription is for the government, or regulatory authority, to identify the party or activity responsible for the externality and to implement corrective policies focused on the apparent source of the problem – the "culprit". One objection to this proposal is that regulatory intervention requires considerable information about the costs and benefits of the contending parties. In theory, a tax could be levied on your local cycling club, the proceeds to be directed back to the residents of Guangzhou – obviously impractical.

In the real world, furthermore, externalities can be a convenient tool of political gamesmanship. In Sydney, Australia, the international airport imposes a curfew on aircraft movements between 11 pm and 6 am. The flow-on impact of this local policy imposes limits on aircraft arrival and departure times on every continent and all major countries and leads also to airport congestion for aircraft, passengers and taxis in Sydney during the restricted operating hours. While a peaceful night's sleep is doubtless highly valued by all Sydney residents, effectively a small number of people – specifically a few hundred thousand voters resident in the electoral constituencies adjacent to the airport – are imposing substantial costs on domestic and international air travellers as well as airlines schedules and air traffic authorities across the globe. Is this the most efficient outcome? Probably not, but the costs and complexities of international negotiations to achieve a better outcome have up to now defied politicians and regulators.

The Coase approach – let the market decide

Sometimes the source of an externality is easy to identify, but the offending party, less so. *Sturges v. Bridgman* is a legal case of 1879 vintage involving a confectioner operating a business in Wigmore Street, London.[7] At the time one of the two large machines used in the business had been in position for more than 60 years, the other for 26 years. A medical doctor arrived to occupy adjacent premises in nearby Wimpole Street and eight years after his arrival he decided to extend his consulting room so that it abutted the confectioner's premises. The doctor then found that the noise of the confectioner's machinery interfered with his medical practice in the newly renovated consulting room.

A court case ensued.

The court found in the doctor's favour. Apparently in this case the so-called "Doctrine of Lost Grant" which holds that "if a legal right is proved to have existed and been exercised for a number of years the law ought to presume that it had legal origin" did not apply – though it has in other, apparently not dissimilar, cases.

Cases involving prior occupancy are, of course, common. Locating an airport near a residential area may seem to be a clear case for compensation to existing residents from an equity perspective. But the case is less clear when applied to residents who move into the area after the airport has been established – or even after the long-term plan to establish an airport had been published. We have an intuitive sense, not necessarily incorrectly, that "first come, first served" (or as Australian ANZAC soldiers might have said, "first in, best dressed") conveys some widely accepted if not formal claim. The British devotion to queuing and rationing in preference to auctions as a basis for allocating scarce but under-priced resources during and even long after wars have been won is testimony to this belief.

Citing selected UK and US court cases, where judges appear to have struggled to find a consistent legal foothold, R H Coase's perspective on externality problems differs from Pigou. For Coase, the socially optimal rate of production is unchanged at OQ_2 in Figure 8.1. However, Coase argues, externalities are a *mutual problem.* The object is to achieve an *efficient* outcome, not to decide which party is at fault and apply financial rewards or penalties and quantitative controls. Failure to understand this has led to diverse and apparently inconsistent outcomes in a pointless legal game of "find the villain."

The *Coase Theorem*[8] states that where externalities exist, providing the property rights are clearly defined but regardless of the initial allocation of property rights, the parties will in principle be able to negotiate an efficient outcome (equivalent to OQ_2 in Figure 8.1). And the outcome will be independent of the initial allocation of property rights.

Like many economic principles, the Coase theorem can be explained with a simple parable.

I am dining alone in a non-smoking restaurant.

The couple at the adjoining table lean over and say: "Do you mind if we each smoke a cigarette?'

I say, "Yes I do mind" and point to the No Smoking sign.

The gentleman says:

"If I pay you £10 cash, will you consent to our each smoking one cigarette?"

"No!" I say.

"£20?"

"No."

"£24?"

"No."

"£25?"

"OK," I say, "but only one cigarette each – and cash payment in advance."

Clearly, I and my fellow diners at the adjacent table are both better off as a result of this voluntary transaction. I obviously value the discomfort of smoke from two cigarettes at £25 or more. The diners on the adjoining table value the enjoyment of their cigarettes at £25 or less. The payment of £25 is a voluntary capitalist act between consenting adults. A mutually beneficial outcome has been achieved and the highest value outcome has prevailed.

Now suppose the situation is reversed.

Smoking is allowed in this restaurant.

I find the smoke from the adjoining table disturbing and lean across to my neighbours.

"If I pay you £10 will you douse your cigarettes?"

"No", they say.

"£20 then?"

"No"

"£24?"

"No"

"£25?"

I may be willing to pay more than £25 for clean air, but we know that this is the minimum my neighbours will accept for desisting to smoke.

You get the picture. If I bid up to £100 and still meet a refusal (at which point I desist) it is clear that the pleasure of smoking is worth more than £100 to my fellow diners. Though I have a strong preference for fresh air, in this context it is less than £100 to me.

Again, the highest value outcome has prevailed.

Consider Figure 3.1 again.

- **First,** assume that the firm has the right to produce at the rate OQ_1. To simplify the negotiation process we will assume that nearby residents and the customers are identical. The consumer values the OQ_1th unit at Q_1B (= OP_1). But the cost of that unit, taking account of the additional social cost imposed on the customer, is Q_1A. The customer will in principle be prepared to offer an incentive of up to AB to the firm to limit production to OQ_1 *minus* one unit. Since $Q_1A > Q_1B$ the firm will clearly accept this offer; and for the OQ_1 *minus* second unit also. In total, negotiating unit-by-unit the customer is willing to offer the firm the area EAB to limit production to OQ_2 and the firm would accept this proposal.
- **Second,** assume that the neighbouring householder has the right to clean air. She can, if she chooses, prevent the firm from operating at all. However, outright closure of the factory would be a sub-optimal outcome for the householder. The first item produced costs the firm *ON* but is worth approximately *OC* to the consumer, substantially more than the total (private plus social) cost involved. And the second item is also worth more than the total cost, even taking into account the accompanying cost of the smoke. The firm can negotiate a production deal with the consumer up to a production rate of OQ_2. But no further. Production of an additional unit beyond OQ_2 imposes a cost > Q_2E (by the SMC curve) but is worth less (as shown by the MR curve). The consumer will impose a production limit at this point.

In either case – as long as ownership of the rights to clean air is defined and transferable – we arrive at the most highly valued rate of production, OQ_2.

The Coase analysis seems to solve the externality problem. But implicit in the Coase solution are a number of tough underlying requirements and assumptions:

- **First,** for negotiations to commence the property right must be clearly defined – either smoking is allowed in the restaurant, or it is not.[9]
- **Second,** that right must be alienable or transferable, otherwise negotiations (along the lines of the above parable) cannot take place – something that would indeed be unlikely in the artificial example of the restaurant.
- **Third,** we have not considered the interests of other diners. The Coase analysis assumes zero negotiation costs, an absence of strategic behaviour and the availability of detailed information. Negotiating with the next table may be feasible. Negotiating with all the customers in a crowded restaurant poses practical difficulties. Nor have we considered the possibility of strategic behaviour. The prospect of a £100 cash pay-off is surely an incentive to light up a cigarette at a table adjacent to a non-smoking diner. But while Coasian negotiations make generous assumptions, note also that Pigou assumes away the inevitable costs (including rent-seeking[10]) that accompany intervention by the government or some other regulator.

When the sums involved are relatively small, where the parties have similar bargaining power and the subject of negotiation is not of critical importance, Coase-type negotiations may lead to an acceptable outcome. It is not difficult, however, to think of situations where practical and ethical difficulties might arise.

For example:

> An international chemical company establishes a production facility that produces toxic fumes next door to an aged pensioner with a lung condition. The pensioner might accept an offer of compensation sufficient to allow him to relocate to a healthier area. It is unlikely that the chemical company would accept an offer of a few hundred dollars from the pensioner to re-locate their production facility. In the case of highly valued goods – and life and death are an extreme example of this – the gap between WTAC and WTP[11] can become extremely large, even infinite, and this is where the initial allocation of rights becomes critical.

We are therefore left with a choice of two alternative but imperfect ways to deal with externalities – which approach will work best (and for whom) will depend on the specifics of each case.

Externalities do not, therefore, constitute an automatic case for government intervention along the lines advocated by Pigou – although the Coase argument does not constitute a watertight case against regulatory intervention either. The challenge raised by Coase is to consider alternative mechanisms that will get us closest to the socially optimal or at least politically acceptable outcome at a low cost. Sometimes this will be best achieved by private negotiation. In other cases – for example where negotiation costs are high, as they may be in determining the operating hours of an international airport or the air quality in a crowded restaurant – the case for regulatory intervention or political action merits consideration.

Note that while Pigou and Coase both guide us to the identical optimal level of production, OQ_2 in Figure 8.1, the processes for arriving at this point, and the financial payoffs from the alternative strategies, are markedly different. In the restaurant example the optimal outcome may have been achieved but depending on the restaurant smoking rules I receive cash or I pay cash. Should the (relatively poor) nearby residents be expected to compensate the multinational smoke-producing factory owner for a reduction in the rate of production? This raises questions not of *efficiency* but of *equity*. Economists generally agree at least in principle about matters of efficiency; about equity, there is wide scope for disagreement even, or especially, among economists.

A third alternative…real-world solutions

The Pigou and Coase thought experiments (for this is what they are) advise different responses to the problem of externalities. There is however an attractive third alternative – perhaps the problem has been solved already!

When all else fails, examine the evidence.

Pigou sat at his desk and analysed externalities on the basis of abstract reasoning – a smoke-emitting factory, a traffic-congested highway. He argued (as economists often do) in abstract terms – "…assume you have a can opener."

Another classic example of an externality thought experiment is the lighthouse.

John Stuart Mill[12] argued "no one would build lighthouses from motives of personal self-interest," because they could not collect the fees necessary to cover costs. Mill's conclusion was therefore that "it is a proper office of government to build and maintain lighthouses … since it is impossible that the ships at sea which are benefitted … should be made to pay a toll."

A more recent, but famous, thought experiment to illustrate externalities was devised by the 1977 Nobel laureate in Economics, James Meade (1907–1995).[13] (Analysis of the economics of property rights seems to have spawned a rich field of Nobel laureates.) In Meade's thought experiment bees collect apple nectar to transform into honey, in the process fertilising the apple blossoms. The apple farmer needs the bees to fertilise the trees; the apiarist needs the apple blossoms to produce honey. Since neither party can capture the benefits of the other, Meade concluded that a set of formulae could "show what taxes and subsidies must be imposed"[14] to arrive at the optimal production of apples and honey. Let to itself, Meade argued, an unregulated market would be inefficient, under-producing both apples and honey.

These are fascinating insights, but they are speculations. Many parables about externalities arising from market failure are used to justify regulatory intervention but they often fall apart on closer inspection. Thought experiments are interesting and inexpensive – but they can only go so far. Ronald Coase[15] moved the discussion away from abstract thought experiments and a step closer to the real world by highlighting court cases selected from US and UK legal records. Coase likewise challenged John Stuart Mill's lighthouse example[16] on evidential grounds.

Stimulated by Coase's analysis, subsequent studies of real-world situations have challenged the easy assumption that third-party or government intervention is the optimal, or even necessary, solution to externality problems. Meade's carefully constructed discussion of honey and apples has been explored in detail by Steven Cheung[17] in the state of Washington in the United States, one of the largest apple-growing areas in the world. It seems that markets are sometimes adept at resolving externalities without the direct intervention of government, or at least with limited resort to government assistance. In Cheung's finding "contractual arrangements governing nectar and pollination services are consistent with an efficient allocation of resources."[18]

Aside from Cheung's finding of a negotiated outcome along Coasian lines – which happens more often than A. C. Pigou might have expected – there are other ways to deal with externalities. Establishing, monitoring and enforcing property rights is not costless. Intervention by regulators can come with heavy costs. This is one reason why externalities are prevalent. Many of life's inconveniences and irritations are simply best ignored. Sometimes they can be resolved by a chat over the garden fence. If the busker's Janis Joplin impersonation is skilful enough to attract a crowd, she may score a free cup of coffee from the Kardomah.

A common response to externality problems is to internalise them. Your neighbour might, for example, consider inviting you to join the backyard barbeque. As Cheung demonstrated, the apiarist and the apple farmer arrived spontaneously at mutually beneficial contractual arrangements. The doctor in Wimpole Street might have considered a purchase and lease-back arrangement on the adjacent confectionary premises. Some large-scale residential property developers are able to optimise privacy on small blocks of land by arranging homes so that no window looks directly into the living areas of the house next door, or casts a shadow. In contrast, a residential area based on blocks of land purchased by individual developers would be unlikely to achieve this. Retail shopping landlords are often able to achieve better returns for their malls than strata owners of individual shops because they can manage the operating hours and (importantly) the tenancy mix of the mall to maximise foot traffic, taking account of the socio-economic profile of the neighbourhood.

Indeed, leases in shopping malls are a rich source of study for the internalisation of externalities. Why not locate your bookstore adjacent to the popular coffee shop? Does your rent not reflect this benefit already?

Property externalities – the sky's the limit

Existing property rights are continuously under challenge and in many cases evolve efficiently to meet changing circumstances. For example, despite the English principle established in 1587 that[19] "*cuius est solum, eius est usque ad coelum et ad inferos*" meaning *whoever's is the soil, it is theirs all the way to Heaven and all the way to Hell* aircraft freely fly between Edinburgh and London. But what about low-flying surveillance drones? The law is still evolving on this topic.

Air rights are a case where active markets in externalities have been established in some jurisdictions. On Coasian principle, rights to the air space above a property must first be defined. These rights can then be used to facilitate or constrain future development of that property. Alternatively, these rights can sometimes be sold or transferred to third parties.

In some US cities, the airspace above a property confers developmental rights which can be sold or transferred. Thus, in a dense downtown area, each building in the area may have the right to thirty-five stories of airspace above its own property. In one possible scenario, owners of a heritage building of only three storeys could capitalise on their assets by selling their building for demolition and allowing a thirty-five-story skyscraper to be built in its place. In a different scenario, a skyscraper developer may purchase the unused airspace above the heritage building to develop a taller building at a nearby location, preserving the heritage building. In 2013, Christ Church in New York sold its vertical development rights for a record US$430 per square foot, making more than US$30 million on the sale for the right to build in the space over its building.

> The average price paid in Manhattan for the transfer of air rights – the undeveloped space above a building – rose 47% in 2013 from the previous year, according to a report from Tenantwise Inc., a New York real-estate services and advisory company.[20]

The first air-rights building in the UK was One Embankment Place, a commercial office building for PwC constructed above Charing Cross station and completed in 1991. Other air-rights buildings include: Alban Gate at London Wall; Broadgate in the City of London and the Cannon Place development above Cannon Street Station.[21]

In the city of Sydney, Australia, some heritage buildings – which have limitations on structural changes – are awarded alienable air rights. In compensation for limiting their ability to develop these sites, owners are able to sell the air rights that exist above these buildings. The result is an active market in air rights. In Sydney CBD the market in heritage floor space rose from AU$ 331/sqm in 2013 to an average of AU$ 615/sqm in 2015 for 18 transactions, with prices ranging from AU$ 450 sqm to AU$ 1,300 sqm. Areas transacted ranged from 2.0 sqm to 3562.6 sqm.[22] Of course the market value of air rights, like vacant development sites, incorporates optionality value – and once used to support a new development the optionality value vanishes.[23] So, the market value of air rights can be a leading indicator of future development projects and like other option markets, values can be volatile.

Air rights can also be employed to limit future development. Donald Trump famously bought the air rights around Tiffany's flagship store in New York to prevent other developers from building up above it and obscuring the views from Trump Tower.

In Central Sydney, a market in "air rights" or Heritage Floor Space (HFS) has existed since 1997. The HFS market provides an incentive to conserve and maintain existing heritage buildings. After completion of approved conservation works an award of HFS is registered with the Sydney City Council and this can be sold or transferred to a development that would otherwise contravene existing Floor Space Ratio limits. The sale of HFS is a private transaction between the owner and the prospective buyer – the City acts as the scheme administrator.

Example: In 2013 an application was approved for the award of HFS for 142–144 Pitt Street, Sydney, Australia.

Built circa 1882, this former warehouse is a reminder of the scale and quality of buildings in this part of Sydney. It has historic significance for its association with John Sands, one of Sydney's oldest publishing companies that occupied the building between 1883 and 1905. The building has aesthetic significance for its extraordinary composition of polychrome brickwork, ceramic tile panelling with zigzag cream brick motifs, modelled cement emblems at the first floor level and window joinery with carved barley-twist mullions, forming a harmonious, engaging ensemble that enriches the streetscape.

Property rights – more than just physical

Alchian and Allen predicted back in 1979 that in the future technology would help to resolve externality problems.[24] They were at least partly correct. Modern inventions such as noise-cancelling headphones and anti-barking electric dog collars help to limit noise externalities. Satellite technology on the dashboard advises commuting motorists which traffic-congested roads to avoid on the journey home. A mobile phone text message allows me to time my arrival at the local railway station just as my mother-in-law's train arrives from Swindon, reducing congestion in the car park and domestic tension. In the age of COVID-19 international travellers can home-isolate with body tracking devices.

Investment in "clean" energy is designed to limit carbon dioxide emissions, arguably a negative externality that threatens life on Earth. International strategies to limit carbon dioxide emissions as a response to a negative externality are the most ambitious examples in the history of regulation. Still evolving, international policies to limit carbon dioxide emissions are a complex mixture of compulsion, taxes and subsidies and moral suasion.

While technology can facilitate the war against negative externalities, new sources of externalities have emerged.

When I look up the definition of "externalities" on Google I leave an electronic information trail that is of potential value to Google's marketing department. From their perspective, my search activity is a positive externality. Attempts to manage these search engines (and to extract payment) are in fact an attempt to internalise an externality. Should you pay Google for using its software – maybe Google should be paying you? At the very least the "free" provision of search engines and other social media platforms goes beyond pure altruism. I take no account of the third-party value that I create when I search on Google; nor can I figure out how to capture this value. Of course, sometimes I do receive some payback – after searching for information on car sales on my search engine I may be rewarded the next day with a deluge of information about used car yards in my neighbourhood – this can be helpful. More disconcertingly, after discussing the purchase of a new car at the family dinner table I suddenly find my website pages crowded with car advertisements – has the computer been listening?

More recently, the COVID-19 global pandemic provides fertile grounds for debate about externalities. Predominantly, governments have responded to the pandemic with a Pigouvian interventionist approach designed to directly limit personal interaction by measures such as restrictions on personal travel and curfews. With hindsight, more targeted policies comprising taxes and subsidies to identify sources of infection, encourage vaccination and protect vulnerable people can be imagined. And in fact, new technologies – such as home quarantine trackers and group meeting software – rapidly emerged and proliferated during the course of the pandemic. Policies to exclude unvaccinated people from public areas such as sports arenas and restaurants open the way for Coase-type negotiations. Vaccine "passports" and "mandates" are currently central to public debate about public infectious disease management. The COVID-19 pandemic has provided many case studies in the management (and mismanagement) of externalities as well as a masterclass on the ever-shifting boundary between private and public goods.

Debates about externalities often become ensnared in ethical debates. Banning household coal and wood fires limits smog and protects the health of vulnerable people, but the impact is likely to be regressive. Poor people often rely on coal and wood for warmth, electricity and jobs; wealthy people have many alternatives. For dinner they outsource to home delivery; to keep warm they winter in the Caribbean. The spectre of global climate change represents a major challenge to the analysis and regulation of externalities. Coase and Pigou offer valuable insights and tools, but not ready-made solutions to those venturing into these shark-infested public policy areas.

Notes

1 Property rights vary widely between jurisdictions. In the UK for example the Crown has a claim on any buried treasure discovered on your property. A meteorite landing on your front lawn, however, probably belongs to you. (Daily Telegraph, 17 March 2021, reader's letter). Even rights to water falling freshly from the sky can be problematic – while it may not be groundwater yet, rain does have that implicit potential. In Colorado, United States, for instance, the question of harvesting rainwater was debated for years before a determination was made: property owners can store up to 110 gallons in barrels for outdoor use only.
2 Pigou, A. C., *The Economics of Welfare*, 1st ed. (London: Macmillan, 1920).
3 Coase, R. H., The Problem of Social Cost, *Journal of Law and Economics,* 3 (1960), pp. 1–44.
4 See Chapter 9 for a discussion on the important but often overlooked distinction between the volume of production, Quantity, and the rate of production, Quantity/time period.
5 Note that PP' is a Price line, *not* a Demand (or Average Revenue) line as in a standard monopoly model. The cotton mill is assumed to be a price taker in the fabric market. The prevailing market price (defined as £ per square meter) is set by broad market demand and supply conditions, not by variations in the mill's rate of production, *Q/t.* We are also assuming that the current market price, OP_1, reflects *both* private and social benefits and costs from the consumption of cotton fabric – that is, all *externalities* are on the production side in this discussion.
6 For simplicity we assume that there are no Fixed Costs in this enterprise. If Fixed Costs exist, the area under the MC curve is Total Variable Cost.
7 See Coase (1960) for the Sturges v. Bridgman case and related discussion.
8 Ronald Coase (1910–2013), awarded the Nobel Memorial Prize in Economic Sciences (1991).
9 We ignore here the mandatory No Smoking rule that exists is many jurisdictions. This may be associated with legal liability of the restaurant proprietors for the health of staff for example, itself an example of a broad brush Pigouvian solution to an externality problem that may have more efficient solutions.
10 See below pp. xxxx.
11 See Chapter 7.
12 Mill, J. S. *Principles of Political Economy*, 1848 (London: Pelican Books, 1970), pp. 342–343.
13 Meade, J. E., External Economies and Diseconomies in a Competitive Situation, *Economic Journal*, 54 (1952), p. 54.
14 Meade (1952), p 58.

15 Coase (1960).
16 Coase, R. H., The Lighthouse in Economics, *Journal of Law and Economics*, 17, no. 2 (1974), pp. 357–376.
17 Cheung, N. S. H., The Fable of the Bees: An Economic Investigation, *Journal of Law and Economics*, 16, no. 1 (1973), pp. 11–33.
18 Cheung (1973), p. 15.
19 See Kohlstedt, K. *From Heaven to Hell: Exploring the Odd Vertical Limits of Land Ownership* (2020). https://99percentinvisible.org/article/heaven-hell-exploring-odd-vertical-limits-land-ownership.
20 *Wall Street Journal*, 23 April 2014.
21 Designing Buildings: The Construction WIKI 2020, *Air-rights buildings*. https://www.designingbuildings.co.uk/wiki/Air-rights_buildings.
22 *Sydney Council Heritage Floor Space Update*, December 2015. See also JLL Research: Why Air is Becoming Hot Property, 29 January 2020. https://www.jll.com.au/en/trends-and-insights/cities/why-air-is-becoming-hot-property.
23 See Chapter 5 on optionality values.
24 Alchian, A., and Allen, W. R., *Exchange and Production* (Belmont, CA: Wadsworth Publishing Company, 1979), pp. 91–95.

Bibliography

Alchian, A., and Allen, W. R. 1979. *Exchange and Production.* Belmont, CA: Wadsworth Publishing Company.

Cheung, N. S. H. 1973. The Fable of the Bees: An Economic Investigation. *Journal of Law and Economics,* 16(1), pp. 11–33.

City of Sydney, December 2015. *Sydney Council Heritage Floor Space Update*.

Coase, R. H. 1960. The Problem of Social Cost. *Journal of Law and Economics,* 3(1), pp. 1–44.

Coase, R. H. 1974. The Lighthouse in Economics. *Journal of Law and Economics,* 17(2), pp. 357–376.

Designing Buildings: The Construction Wiki. 2020. *Air-Rights buildings.* [Online] Available at: https://www.designingbuildings.co.uk/wiki/Air-rights_buildings [Accessed 22 July 2022].

JLL Research, 29 January 2020. *Why Air is Becoming Hot Property.* [Online] Available at: https://www.jll.com.au/en/trends-and-insights/cities/why-air-is-becoming-hot-property [Accessed 23 July 2022].

Kohlstedt, K. 2017. *From Heaven to Hell: Exploring the Odd Vertical Limits of Land Ownership.* [Online] Available at: https://99percentinvisible.org/article/heaven-hell-exploring-odd-vertical-limits-land-ownership/ [Accessed 22 July 2022].

Meade, J. E. 1952. External Economies and Diseconomies in a Competitive Situation. *Economic Journal,* 54, p. 54.

Mill, J. S. 1970. *Principles of Political Economy (1848).* London: Pelican Books.

Pigou, A. C. 1920. *The Economics of Welfare.* First ed. London: Macmillan.

9 Natural resources – old problems, changing rules

Confronting "the surly bonds of earth"[1]

The impact of human activity on the global environment now commands international attention. But the debate about the limits of the Earth's resource endowment is not new. The 19th-century founders of the science of economics believed they lived in a period of transient prosperity. Finite natural resources would ultimately limit economic growth, setting a ceiling on population numbers and curtailing the wealth creation process.

In contrast, most current financial and economic analysis (including analysis applied to real estate markets) is based on propositions and prescriptions that emerged early in the 20th century. This is the so-called neo-classical paradigm: future trends are broadly a continuation of the past; economic growth will continue in perpetuity. Thanks to the miracles of expanding scientific knowledge and technical progress the binding tension between a finite nature and limitless wealth creation that dominated the 19th-century analysis has been relaxed, if not in perpetuity then at least indefinitely.

Neoclassical analysis and models comprise a formidable set of tools, compelling policy prescriptions and powerful insights. Nevertheless, the 21st century seems likely to challenge at least some of the implicit underlying assumptions. Trade-offs between short-run profit-maximising and long-run value optimising strategies, often packaged as "sustainability" or Environmental, Social and Governance (ESG), have re-emerged on the agenda for analysts and asset managers. New insights and more flexible models are required to value and manage assets in finite or resource-limited situations. The decision rules and optimisation benchmarks applicable in the neo-classical world, often presented without qualification in economics textbooks, can deliver sub-optimal, even perverse, outcomes in our re-emerging, natural resource-constrained, reality.

Our finite planet – a brief history

Analysis of Earthly limits, in particular the topics of environmental degradation and global climate change, now occupies centre stage in domestic and international policy forums. The real estate sector, from the twin perspectives of land utilisation and construction activity, is often identified as a prime source of environmental

DOI: 10.1201/9781003111931-9

pressures. This would not surprise the early writers in what has become the science of economics. In fact, they might ask: what has taken so long? Most of the writers of the 18th and 19th centuries on the subject then described as "political economy" believed they were living in a period of transient prosperity. Thomas Malthus (1766–1834), sometimes described as the Founder of Economics, was the first to propose what we would recognise as a formal environmental model.

Malthus was a pessimist. According to his model, the population grows as a geometric progression (1,2,4,8,16... = 2^n) whereas food production grows as an arithmetic progression (1,2,3,4...*n*). Inevitably therefore the limits of the latter will impose constraints on the former, leading to falling living standards; ultimately famine will impose a harsh ceiling on population growth.

And indeed, there are real-world examples that appear consistent with this gloomy forecast.[2]

Malthus was not alone. According to John Stuart Mill (1806–1873) writing in 1848:

> It has always been seen, more or less distinctly, by political economists, that the increase of wealth is not boundless: that at the end of what they term the progressive state lies the stationary state, that all progress of wealth is but a postponement of this...[3]

However, the early economists disagreed on the specific mechanism that would close their growth models. For example, Mill postulated a three-factor model – labour, capital and land. In his analysis, the increased pressure on limited natural resources such as land would drive a rise in the cost of living. This in turn would compress the real wages of workers; with falling demand for their products, falling profits would limit the returns to capital. The long-term winners, Mill believed, would be the landlords.[4]

Even those economists who modelled a world of limited productive resources – such as David Ricardo (1772–1823) who also identified arable land as the binding constraint[5] – made assumptions about other resources being abundant and therefore free:

> ...the brewer, the distiller, the dyer, make incessant use of the air and water for the production of their commodities; but as the supply is boundless, they bear no price.[6]

Ricardo assumed (as we might not) that clean water and fresh air are limitless, and therefore free, resources.

For many of these early economists, as for some present-day commentators, beyond the current window of transient prosperity, the future of mankind was a world of "shallows and miseries."[7]

The news was not all bad, however. Karl Marx, a prominent exponent of an apocalyptic end to the capitalist growth model, looked forward with some enthusiasm to the post-capitalist stage in mankind's development. This would involve a world where:

> ... nobody has one exclusive sphere of activity but each can become accomplished in any branch he wishes, society regulates the general production and thus makes it possible for me to do one thing today and another tomorrow, to hunt in the morning, fish in the afternoon, rear cattle in the evening, criticise [i.e. philosophise] after dinner...[8]

Not perhaps a lifestyle that appeals to us all. But even Mill found consolation in a limited growth model. In contrast to his own world, where, as he observed:

> The normal state of human beings is that of struggling to get on; that the trampling, crushing, elbowing and treading on each other's heels, which form the existing type of social life... (p. 113).

For Mill, some future, more leisured, less competitive, society had many attractions.

In a more recent era, John Maynard Keynes (1883–1946) thought so, too. With rising affluence and productivity, Keynes envisaged a world of:

> ...three hour shifts or a fifteen hour week.[9]

Most of these writers, we should observe, were of private means or in receipt of a comfortable income. Independent of the immediate need for regular work and a weekly wage, leisure offered an attractive alternative to those destined for a lifetime of toil. Keynes even quoted a traditional epitaph of a mythical charwoman:

> Don't mourn for me, friends, don't weep for me never,
> For I'm going to do nothing for ever and ever.[10]

Regardless of the end state, all these writers anticipated a future state of the world, nasty or nice, where economic expansion would grind to a halt.

There is however a second tradition in the history of economic thought, a tradition to which financial analysts and real estate professionals (usually inadvertently) subscribe.

By the end of the 19th century, it was becoming clear that the early economic models were not working out as planned. Population growth remained high, but food production outstripped it. Life expectancy and living standards rose. Technology and science, rail and sea transport and global trade seemed, at least for the foreseeable future, to have eased the binding limits previously imposed by limitless population growth and limited land productivity. Might economic growth, indeed, be boundless? Keynes indeed proclaimed:

> ...that mankind is solving its economic problem.[11]

Economic theorists in the early 20th century, therefore, relaxed the constraints of food availability pitted against population growth, that so worried classical economists. Ever taller buildings, canals and railways relieved pressure on urban land availability; mass production, refrigeration and new technology (the railroads, the telegraph, steamships and the telephone in particular) eased, at least temporarily, the tyranny of distance. Accordingly, the economic problem was redefined. The focus of economic and financial analysis turned to policies and analysis designed to optimise rates of production and consumption, maximising profits, minimising under-employed resources and, implicitly, maximising Present Value (*PV*). The future would replicate the growth and wealth creation of the recent past and in perpetuity. The Golden Age had finally arrived.

Once *time* is removed from the analysis as an explicit variable and becomes a mere parameter, the way is open for the neo-classical analysis that forms the core of modern microeconomics and financial analysis. Optimisation is the core objective – whether company profits, lifetime wealth creation or personal utility management. Modern microeconomics texts often deal with the variable of *time* as a purely technical matter – in a way in which the classical scholars did not.

This is a view that economists accept when they prescribe profit-maximising conditions for enterprises and natural rates of interest. It is implicit when property analysts close their discounted cash flow (DCF) valuation models by capitalising future income. Analysts and investors may think that real estate valuation is a value-free, technical subject. But as Lord Keynes warned:

> Practical men, who believe themselves to be quite exempt from any intellectual influences, are usually the slaves of some defunct economist.[12]

The 1973 global oil crisis, the discovery of a hole in the Antarctic ozone layer (mid-1980s) and more recent debates about the existence of, possible causes and responses to, global climate change are a sobering reminder that a broad perspective on long-term resource utilisation and wealth creation is sometimes required. These concerns are not limited to global topics. Environmental pressures and the conservation of scarce resources increasingly challenge the orthodox microeconomics, time-agnostic, optimisation criteria that guide business decisions at the individual asset or enterprise level.

As we rediscover a world of depletable, and sometimes exhaustible, resources, the future is not a simple extrapolation of the past. Decisions that seem to be optimal or profit-maximising at a point in time may not be consistent with long-term optimising strategies. Capitalising rental income after, say, ten years, is a convenient way to close a discounted cash flow (DCF) model for a single real estate asset. But this calculation implies rigid assumptions about the future.[13] In many cases the exploration of alternative terminal conditions for a financial model is instructive and can produce very different outcomes, investment strategies and policy prescriptions.

Demand and supply – read the small print!

Every introductory microeconomics text introduces a market model with a supply/demand diagram. In the frictionless market (with no transaction costs) represented in Figure 9.1 the equilibrium or market-clearing quantity is Q_e and the equilibrium price is P_e.

But assume this diagram describes a residential dwelling market. What "quantity" are we measuring on the horizontal axis? Are we measuring the *stock of existing dwellings* at a point in time, *t*? Or, as shown in Figure 9.1, the *rate of construction of new dwellings per time period* at time *t*? Alfred Marshall (1842–1924), an early sponsor of the supply/demand market construct, was very careful to define the horizontal or quantity axis as a stock, not a flow, and as a representation of a specific market at a point in time. Later writers are less careful. And many textbooks are silent, or at least ambiguous, on this point.[14]

In a world where resources are limited – for example, *exhaustible resources* such as minerals and fossil fuels, or *depletable but renewable resources* such as fish and timber – the distinction between the *stock* at time *t*, $Q(t)$, and the *rate of change of stock* at time *t*, the flow, $q(t)$, where we differentiate the stock with respect to time to establish the rate of flow:

$$d\left(Q(t)\right)/dt = q(t)$$

is all important. The next section demonstrates why this is so.

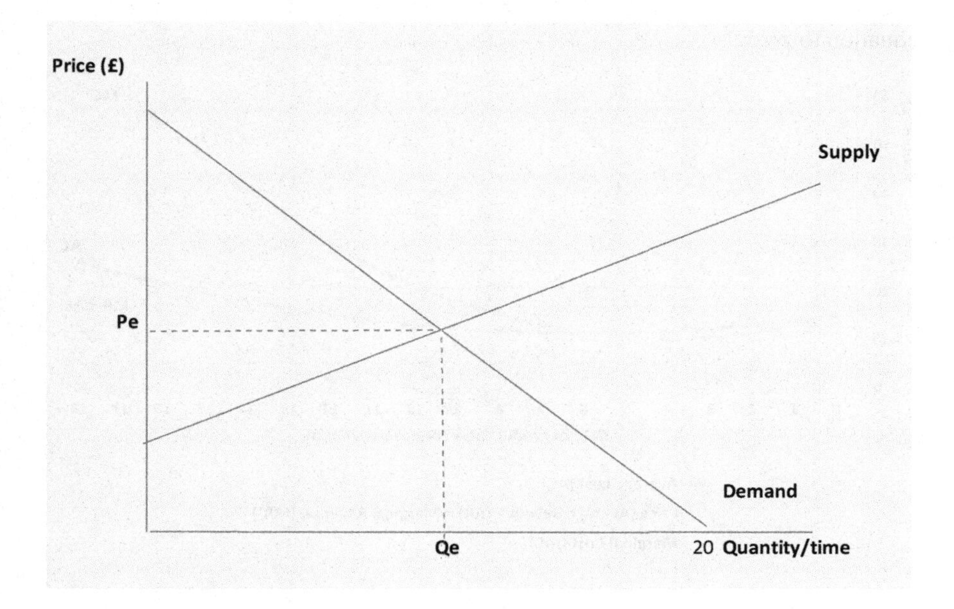

Figure 9.1 The Demand/Supply Market Model

Back to the future – time for a change?

Consider a standard microeconomic textbook optimisation problem.

Figure 9.2 illustrates the standard textbook model of an enterprise in a state of perfect competition. The market price of the output is £21 per unit regardless of the rate of production – this enterprise is a price taker. Since the unit price is assumed constant at £21 it follows that:

$$\text{average revenue per unit }(\text{AR}) = \text{marginal revenue per unit }(\text{MR}) = £21.$$

The rising Marginal Cost (MC) and Average Cost (AC) curves as the rate of production increases reflect the conventional engineering assumption of diminishing returns In Figure 9.2, AC is a minimum (£11.40) at a production rate of 6 tonnes per year. The MR and MC curves intersect at a production rate of 10 tonnes per annum. At this production rate AC = £13.00.[15]

The optimising problem to be solved by the owner of a productive enterprise is to maximise profit by adjusting the rate of production, $\prod(q)$

$$\text{Profit} \prod(q) = TR(q) - TC(q) \tag{9.1}$$

where $TR(q)$ is the total revenue for a rate of production q, and TC is Total Cost for production rate q.

Textbook analysis advises that profit is maximised at the production rate q^*. The rate q^* is established when we differentiate Eq (9.1) with respect to q and set the equation to zero:[16]

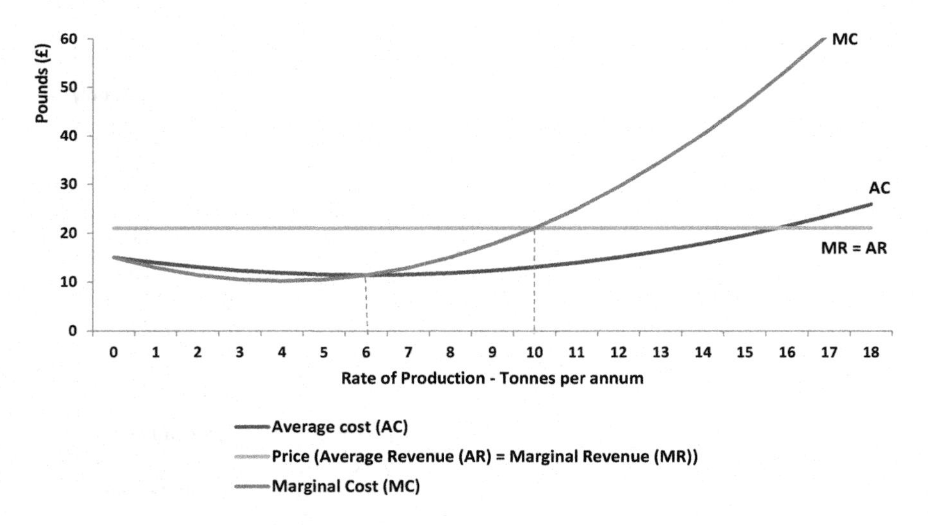

Figure 9.2 The Perfect Competition Model

$$\Pi'(q^*) = TR'(q^*) - TC'(q^*) = 0$$

or

$$\text{Marginal revenue}(\text{MR}) = \text{Marginal cost}(\text{MC}).$$

This establishes the so-called first-order condition for profit maximisation.

In Figure 9.2 this is a production rate of 10 units per year and an annual profit of:

$$10 \times (£21.00 - £13.00) = £80.00$$

If we assume a discount rate, *d*, of 5 percent per annum in a world without inflation then, if the next annual payment is one year hence,[17] capitalising the enterprise income in perpetuity, the *PV* value of the enterprise is:

$$PV = £80.00/0.05 = £1{,}600.00$$

Once this is established, the textbook analysis stops. No further management action is required.

The implicit (but rarely specified) assumption in the textbook model is that this enterprise will operate in perpetuity. This may be a reasonable assumption for say a fruit juice bottler, a T-shirt manufacturer or a coastal desalination plant. But now let's add some additional context to this model.

Suppose Figure 9.2 represents a mining operation. Let's assume a finite ore body of 120 tonnes.

Now, what is the optimal rate of production?

At a rate of production of 10 tonnes per annum, where MR = MC, the mine will be exhausted in 12 years. Profit will be £80.00 per year. The value of the mining enterprise, *PV*, depends on the discount rate, *d*. Let's assume *d* = 5% p.a.

Then

$$PV = \frac{80}{(1.05)} + \frac{80}{(1.05)^2} + \ldots + \frac{80}{(1.05)^{12}}$$
$$= £709.06$$

But is this the *PV*-maximising strategy as we conduct our analysis today?

Suppose instead we choose a production rate of 8 tonnes per annum. AC per unit = £11.80. The mine life is now 15 years (8 tonnes × 15 years = 120 tonnes), and the annual profit is now:

$$8\,x\,(£21.00 - £11.80) = £73.60$$

The enterprise value is

$$PV = \frac{73.60}{(1.05)} + \frac{73.60}{(1.05)^2} + \ldots + \frac{73.60}{(1.05)^{15}}$$

$$= £763.94$$

Evidently, the MR = MC rule, prescribing a production rate of 10 tonnes per year, does not maximise the *PV* of the enterprise in this case, though it does maximise the profit each year during 12-year the life of the mine.

Consider, alternatively, a production rate of 6 tonnes per annum. This production rate maximises *profit per unit*. The mine life is now 20 years (6 tonnes per year x 20 years = 120 tonnes), AC per unit is at a minimum, £11.40, and annual profit is £57.60:

$$PV = \frac{57.60}{(1.05)} + \frac{57.60}{(1.05)^2} + \ldots + \frac{57.60}{(1.05)^{20}}$$

$$= £717.82$$

Evidently, in this example, *PV* maximisation is achieved at neither the *maximum profit per unit* nor the *maximum profit per year* rate of production.

The core to this problem is a trade-off between maximising *profit per time period* – implicitly the objective of the standard textbook MR = MC advice, which implies a production rate q = 10 tonnes p.a. – and maximising *profit per tonne* which implies q = 6 tonnes p.a. As Figure 9.2 shows, the *slower* the rate of production, down to 6 tonnes per annum, the *higher the profit per tonne.* But a lower rate of production comes with an opportunity cost, defined by the discount rate *d.* As long as $d > 0.0\%$ it remains the case that, all else equal, money now is worth more than money later.

It is worth noting that had we assumed $d = 0.0\%$, PV = £960.00 at a production rate of 10 tonnes per annum and £1,140.00 at an 8 tonnes per annum rate. However, the *PV* maximising production rate when $d = 0.0\%$ would be 6 tonnes per annum, when AC = £11.40. This production rate, q = 6, maximises *profit per tonne,* not *profit per year,* and PV = £1,152.00 (= 120 tonnes × £9.60).

So, what is the answer to the mining problem – or to similar problems where the life of an enterprise or a machine or a resource is finite? These are not rare real-world problems.

Let's consider two cases:

Case 1: Optimal strategy – fixed production technology, finite resource

Assume Figure 9.2 describes a mining enterprise with a finite resource (120 tonnes) and $d > 0.0\%$. The extractive technology is defined by the AC and MC curves. The enterprise is a price taker at a fixed price of £21.00 per tonne. Once the mine is designed and constructed, the rate of production, q^*, is assumed to be fixed for the life of the mine. The task is to design the mine with a fixed rate of production, q, which maximises PV.

Recall that the mine owner faces a trade-off between maximising *profit per tonne* (q = 6.0) and *profit per year* ($q = 10$).

We have already illustrated that both rules

$$MR(q^*) = MC(q^*) = 10 \text{ tonnes per annum}$$

and

$$q^* = \text{minimum } AC = 6 \text{ tonnes per annum}$$

are suboptimal for a positive discount rate. In neither case does this deliver the maximum net present value (PV) when

$$0 < d < \infty$$

The problem we face is to select a value of $q = q^*$ that maximises

$$PV = \frac{\Pi(q^*)}{(1+d)} + \frac{\Pi(q^*)}{(1+d)^2} + \ldots\ldots + \frac{\Pi(q^*)}{(1+d)^T}$$

Since the ore body is finite (120 tonnes) and production will be at a fixed rate, q^*, it follows that the life of the enterprise, T, is

$$T = 120 / q^*$$

We can establish the profit-maximising rate of production, q^* as the solution to[18]

$$\{\Pi'(q^*)[(1 - e^{-rT})/d]\} - \{\Pi(q^*)\,[e^{-rT} \,.\, T/q^*]\} = 0 \qquad (9.2)$$

Table 9.1 Optimal production rates (tonnes per annum) for selected discount rates (*d*)

Optimal Production Rates (tonnes p.a.) For Discount Rates (d)						
d	0.0%	0.1%	1.0%	5.0%	10.0%	20.0%
q*	6.00	6.08	6.63	7.93	8.72	9.47
T	20.0	19.7	18.1	15.1	13.8	12.7

Table 9.1 sets out the *PV*-maximising production rate for alternative values of the discount rate, *d*. When $d = 0.0\%$ there is no financial penalty for postponing production. Indeed, the owner could mothball the mine indefinitely without financial cost. The task is simply to maximise profit per tonne of ore extracted, which is achieved at $q^* = 6.0$ tonnes p.a. At this rate of production, the mine life is 20 years. For very low discount rates (e.g., $d = 0.1\%$) the optimal rate of production is slightly faster ($q^* = 6.08$ tonnes p.a.) reflecting the (modest) opportunity cost of delayed production. As *d* rises, the *PV*-maximising production rate increases towards 10 tonnes per annum, the rate that maximises profit per time period. Higher rate of q^* of course reduces the life of the mine, *T*. The textbook prescription, MR = MC and $q^* = 10$ tonnes p.a. and a mine life of 12 years is an extreme solution.

The appropriate discount rate may be determined by market forces. But higher business, political, market or regulatory risks also lead to higher discount rates applied by a typical enterprise. Eq (9.2) shows that the response is faster production and earlier exhaustion of the mine.

The rate of production is not the only policy lever available to the enterprise owner. In this analysis, we have implicitly assumed that the ore body is of uniform grade. In some mining operations, this may not be so. Even if production technology is fixed so that the rate of tonnes extracted, q^*, is constant, as assumed here, there may be an option to accelerate income by selecting higher grade ore. The impact of a rise in *d* is the same – even if the rate of ore extracted is fixed, a policy of "high-grading" a mine brings the cash flow forward; the mine life is reduced, and the total volume of ore extracted may also be reduced.

Case 2: Optimal strategy – flexible production technology, finite resource

Assume now that we can vary the extraction rate, *q*, over time. We now define the optimal (or *PV*-maximising) production rate not as a constant, q^*, but as a path, $q^*(t)$, which can vary through the life of the mine. Consider the mining problem set out in Figure 9.2 a little further.

Let's conduct a mental experiment:

First, suppose that the ore body is very large indeed – infinitely large in fact. This is the implicit assumption of the standard textbook analysis which advises the MR = MC and $q^* = 10$ tonnes per annum production strategy in a timeless world. The only limit on the rate of production is the engineering constraint of diminishing returns, defined by the upward sloping cost curves. Two of the essential ingredients to most real world economic problems, the limits of nature's bounty and the opportunity cost of time, d, have been defined away.
Second, reverse the mental experiment. Suppose you have a finite ore body and have mined it out. You are about to extract the last (120th) tonne of ore. At what *rate of production* will you extract the *final tonne of ore*?

The answer is to extract the final, 120th, tonne at a *production rate* of 6 tonnes per annum. This yields the maximum profit, £9.60, on the last tonne of ore. Since this is the terminal period in the life of the mine the discount rate, d, is irrelevant. Extraction at the prescribed MC = MR rate of 10 tonnes p.a. yields a profit of only a rate of only £8.00 on the last tonne of ore. Applying these intuitions, if you have a very large body of ore and $d > 0.0\%$ the PV-maximising rate of production is initially slightly less than 10 tonnes per annum. As the ore body is mined out the PV-maximising rate of production declines until, as you mine the final tonne of ore, the optimal rate of production falls to 6 tonnes per annum.

The problem is now to maximise PV where

$$PV = \frac{\Pi(q(1))}{(1+d)} + \frac{\Pi(q(2))}{(1+d)^2} + ... + \frac{\Pi(q(T))}{(1+d)^T} \tag{9.3}$$

The PV-maximising rate of production (now defined as $q^*(\mathrm{t})$) is allowed to vary through time, in contrast to the fixed rate q^* in Case 1. This implies also that T is also free to vary, although total production, $Q(T)$, remains fixed by the mine reserve:

$$Q(T) = \sum_{t=0}^{T} q(t) \ dt = 120 \text{ tonnes}$$

The solution to Eq (9.3) involves dynamic optimisation techniques which extend beyond the scope of this text.[19] We can however simulate a set of solutions using the Excel Solver function. The optimal output is set out in Table 9.2.[20]

Total tonnes to be extracted, $Q(T)$, is fixed at 120 tonnes in all scenarios. Each scenario is defined by a different discount rate, d. The only adjustable

Table 9.2 Optimal production rates – flexible technology

Discount Rate (d)	*0.0%*	*2.0%*	*5.0%*	*10.0%*	*15.0%*	*25.0%*
PV (£)	1152.00	958.76	759.81	568.13	444.20	300.47
Year	< ----------------------		-Tonnes per annum-			---------------------- >
1	6.0	7.4	9.0	9.4	9.6	9.9
2	6.0	7.3	9.0	9.3	9.5	9.8
3	6.0	7.2	8.9	9.2	9.4	9.8
4	6.0	7.1	8.9	9.1	9.3	9.7
5	6.0	7.0	8.8	9.0	9.2	9.7
6	6.0	6.9	8.7	8.9	9.1	9.6
7	6.0	6.9	8.7	8.8	9.0	9.5
8	6.0	6.8	8.6	8.7	8.8	9.3
9	6.0	6.6	8.5	8.5	8.6	9.2
10	6.0	6.5	8.4	8.3	8.3	8.9
11	6.0	6.4	8.3	8.1	8.0	8.6
12	6.0	6.3	8.2	7.9	7.6	8.2
13	6.0	6.2	8.1	7.6	7.1	7.7
14	6.0	6.0	8.0	7.3	6.4	
15	6.0	5.9				
16	6.0	5.6				
17	6.0	5.4				
18	6.0	5.2				
19	6.0	3.3				
20	6.0					
Total tonnes (*Q*)	120.00	120.00	120.00	120.00	120.00	120.00

lever for the manager is the rate of extraction, $q(t)$. Note in Table 9.2 that as d rises the *PV*-maximising rate of production $q^*(t)$ also increases, but inevitably the life of the mine contracts. Note also, that for any $d > 0.0\%$ the optimal rate of production declines through time – this reflects the trade-off between maximising *profit per time period* and *profit per unit* of ore. When d is very high ($d = 25\%$, for example) the initial optimal rate of production is close to 10 tonnes per annum, the maximum profit per time period. When $d = 0$ the timing of production is immaterial. Therefore, in this limiting case, PV is maximised when $q = 6$, the maximum profit per tonne and the highest *PV*.

Evidently in the case of an exhaustible resource – such as a mining operation – and for $d > 0\%$, the optimal (*PV*-maximising) rate of production is not 10 tonnes per year. Nor is it 6 tonnes per year.

The essential point is that the textbook rules for maximising profit (or *PV*) require modification for renewable or exhaustible natural resources. Machines and buildings are examples of assets that are both renewable and exhaustible. Therefore, a similar analysis applies to many problems we confront in property markets.

Outcomes can appear counter-intuitive

- It has been assumed in this illustration that the volume of ore (or rock) extracted is synonymous with the volume of refined product (e.g., copper) produced. This will be correct if the ore body is of uniform grade. In the real world, this is seldom the case. If we adopt the conventional assumption that the mine owner's objective is to maximise the *PV* of the project, a rise in the price of copper, if regarded as *permanent and for* $d > 0.0\%$, will *lower* the cut-off grade (the lowest grade of ore extracted) leading to the exploitation of lower grade ore and, if the rate of production (tonnes of rock processed per time period) is fixed, this implies a *reduction* in the current output of copper – though over the extended life of the enterprise, the total volume extracted will be *increased,* as will be the *PV* today. The supply curve of refined copper from this mine is therefore "backward sloping." Conversely, if the rise in the copper price is believed to be *temporary,* the reverse applies – the optimal (or *PV* maximising) strategy may be to temporarily "high grade" the mine, maximising current copper output and profit while the price is high, even though the long-term result may be a reduction in total ore extracted over the life of the mine. The precise trade-off will depend on the prevailing discount rate, *d,* the geological and cost profile of the individual mining operation and whether the price rise is judged to be permanent or transient.
- A policy proposal to terminate the mining of coal, perhaps for environmental reasons in, let's say, the year 2050, will likely lead to high grading of existing mining operations and a temporary increase in the supply of coal, albeit at a lower price, leading to a temporary increase in coal production as well as, perhaps, stockpiling and a higher rate of consumption. Prospecting for new coal reserves will likely decline, although funding for research into alternatives to coal will probably increase.
- Policies to preserve or limit resource exploitation can have perverse effects. For example, policies to limit fishing days or licences to limit the number of trawlers, designed to protect fish stocks, will likely result in investment in larger, more powerful fishing trawlers and the use of sonar and other technologies to seek out fish shoals. Pressure to limit access to capital for mining or renewable resource operations will raise the investment hurdle rate, increasing the focus on high-grade resources with shorter mine life profiles. The financial incentive to conserve resources for the future – or in the case of renewable resources, to harvest on a sustainable basis – may be reduced.

The environmental challenge: Nuclear versus solar: Which is "cheaper"?

The previous section demonstrates that the optimal termination date of a machine (or a durable asset such as a building) is a complex function of financial factors and the rate of physical depreciation of the existing asset. Embedded in these calculations is the discount rate, *d.*

However, a simpler model than Eq (9.15) can offer some insights into the choice between alternative projects or technologies. Consider, for example, the choice between nuclear and solar energy options to supply electricity.

Consider an electricity generating project – it could be either nuclear or solar energy.

$$PV_0 = I_0 + \frac{C}{(1+d)} + \frac{C}{(1+d)^2} + \ldots + \frac{C}{(1+d)^T} \tag{9.4}$$

where I_0 is the initial investment and C is the annual operating cost. The construction takes one year, and the plant begins operating in Year 1. The discount rate is d, and the project has an operating life of T years, after which it must be demolished and replaced by an identical generating plant.

Applying the algebra in Appendix 5A, Eq (9.4) simplifies to

$$PV_0 = I_0 + \frac{C}{d}\left(1 + \frac{1}{(1+d)^T}\right)$$

Since the construction of a new plant takes one year by assumption, construction of the replacement plant must begin in Year T so that production can commence in Year $T + 1$ as the previous plant is decommissioned. Therefore, the cash flow for the replacement plant is

$$PV_1 = \frac{I_0}{(1+d)^T} + \frac{C}{(1+d)^{T+1}} + \ldots + \frac{C}{(1+d)^{2T}} \tag{9.5}$$

And for the second replacement plant

$$PV_2 = \frac{I_0}{(1+d)^{2T}} + \frac{C}{(1+d)^{2T+1}} + \ldots + \frac{C}{(1+d)^{3T}} \tag{9.6}$$

Adding the perpetual chain of plants:

$$PV = PV_0 + PV_1 + PV_2 + \ldots \infty \tag{9.7}$$

Multiplying both sides by $1/(1+d)^T$

$$PV\left(1/(1+d)^T\right) = PV_1 + PV_2 + \ldots \infty \tag{9.8}$$

Subtracting Eq (9.8) from Eq (9.7) and applying the algebra from Appendix 5A we can re-write Eq (9.7) as

$$PV - PV(1/(1+d)^T) = PV_0$$

Or

$$PV = ((1+d)^T)/(1-(1+d)^T)\ PV_0 \tag{9.9}$$

Table 9.3 Alternative technologies – cost comparison

	Inputs		*Solar*	*Nuclear*
Initial investment	Io	£	100	200
Operating cost	C	£ per annum	5	6
Discount rate	d	% per annum	5.0%	5.0%
Operating life	n	Years	15	40
Present value – first plant	PV_0	£	248.10	337.05
Present value – all plants	PV	£	478.05	392.85

Table 9.3 sets out hypothetical data for two technologies – nuclear and solar.

Solar energy has a lower upfront cost (I_0) than nuclear (£100 versus $200) and a lower operating cost (£5 per unit versus £6 per unit). But the life expectancy of the plant is 15 years compared to 40 years for nuclear energy. Applying a discount rate of 5 percent p.a., the cost of the solar option is £248.10, a less costly option than nuclear (£337.05) for the life of one plant. But taking account of a perpetual flow of future plants the nuclear option is less costly (£392.85 compared to £478.85).

So, which is the less costly option?

Note that the ranking between solar and nuclear energy depends on the discount rate. At a discount rate of 6.90 percent p.a. the *PV* of the two options (measured across a perpetual number of future plants) is identical (£314.81).

Note that we are making no assumptions about future technology. So, our analysis assumes an infinite sequence of plants of identical technology. The final variable is *d*, the discount rate which, we assume, reflects broad economic factors such as economic growth and sovereign risk and is therefore identical for all technologies.

Machines and buildings – optimal replacement policies

In the previous section, we assumed finite lives for solar and nuclear electricity-generating plants. But for durable assets (such as machines and buildings) production lives are seldom so easily defined. In contrast to an exhaustible resource such as a mining operation, or a renewable (but potentially exhaustible) resource such as a fish stock, machines and buildings have potentially infinite lives. The limitations are likely to be technical – a more efficient machine may be designed for example – or financial – operating and maintenance costs may rise over time and with accumulated use. It may be that regulatory requirements – e.g., a requirement to use unleaded fuel – terminate the use of a machine.

Consider a machine or a building with the following characteristics:

$$\Pi(q,t) = pq(t) - C(q(t)) - M((q(t), t)) \tag{9.10}$$

We might think of this as a taxi, for example. Revenue is directly proportional to the distance travelled, $q(t)$. Running Cost, $C(q(t))$, is a function of distance and Maintenance Cost (M), is a function of distance and time.

PV is defined by

$$PV = \int_0^T e^{-dt}\, \Pi(q,t)\, dt$$

where d is the discount rate applicable to the asset. Note that in contrast to the mining example there is no constraint on quantity $\sum_{t=0}^{T} q(t)$. Increased output (or distance travelled) in period 1, $q(1)$, does not necessarily imply reduced output in a subsequent period, or a shorter terminal time, T. It follows that the PV-maximising output can be established independently for each time period without discounting.

Differentiating Eq (9.10) with respect to $q(t)$ and setting the derivative equal to zero to establish the first-order profit-maximising condition.

$$\partial\Pi(q,t)/\partial q_t = p - dC_t/dq_t - \partial M_t/\partial q_t = 0$$

Therefore

$$p = dC_t / dq_t + \partial M_t / \partial q_t \tag{9.11}$$

The first order optimising condition is that price (or marginal revenue) is equal to the sum of two terms: running and maintenance MC.[21]

Assuming we can solve Eq (9.11) for the profit-maximising rate $q^*(t)$, the optimal or PV-maximising rate of operating the machine is:

$$PV_1 = \int_0^T e^{-dt}\, \Pi(q^*,t)\, dt - Io + S(T)e^{-dT} \tag{9.12}$$

Io is the initial cost of the machine at $t = 0$. $S(T)$ is the scrap value of the machine at time T.

We establish the optimal terminal time, T, by differentiating Eq (9.12) with respect to T and setting the expression equal to zero.

$$d(PV_0)/dT = \left(\Pi(q^*,T) - dS(T) + S'(T)\right)e^{-dT} = 0$$

or

$$\Pi(q^*,T) + S'(T)) = dS(T) \tag{9.13}$$

The first order condition for a maximum is that at a terminal time, T, the profit from the machine plus the depreciation cost, $S'(T)$, is equal to the interest forgone by scrapping the machine, $dS(T)$. Since profit is a declining function of time it is evident that a higher rate of discount, d, will bring forward the decision to scrap the machine. Higher discount rates reduce the life of durable assets such as machines and buildings. The life cycle of the machine is determined by engineering as well as financial parameters.[22]

Note that the choice here is between operating and scrapping the machine. In the case of taxi owner/driver for example, the choice analysed here implies that at time T the taxi owner will sell the cab and invest the returns at rate d, retiring permanently from the industry.

A more complex calculation is required to establish the optimal time to replace one cab with another new cab – this is a "chain of machines" problem.[23] This is the problem faced by, for example, the operator of a car fleet. This is relevant to the obsolete building problem too. Typically, an old building is demolished to be replaced by another building. What is the optimal life of an office building?

To analyse the "chain of machines" problem we define the first machine as Eq (9.12), and subsequent machines as follows:

For the second and third machines:

$$PV_2 = \int_T^{2T} e^{-dt}\,\Pi(q^*,t)\,dt - I_0 + S(T)e^{-d2T} = PV_1 e^{-d1T}$$

$$PV_3 = \int_{2T}^{3T} e^{-dt}\,\Pi(q^*,t)\,dt - I_0 + S(T)e^{-d3T} = PV_1 e^{-d2T}$$

And for machine k,

$$PV_k = PV_1 \cdot e^{-d(k-1)T}$$

Then the PV for an infinite chain of machines can be expressed as:

$$PV = \sum_{k=1}^{\infty} PV_k = \frac{\int_0^T e^{-dt}\,\Pi(q^*,t)\,dt - I_0 + S(T)e^{-dT}}{1 - e^{-dT}} \tag{9.13}$$

Differentiating with respect to T, we establish that:

$$\Pi(q^*,T)+S'(T)=(1/w)\left[\int_0^T e^{-dt}\,\Pi(q^*,t)\,dt-I_0+S(T)\right] \tag{9.15}$$

and

$$w=(1-e^{-dT})/d$$

where w is the PV of a one-dollar income stream discounted at rate d for T years.[24] Eq (9.15) and therefore w are calculated in continuous time. A discrete-time (e.g., monthly, quarterly or annual) approximation of w is

$$w=\frac{£1.00}{(1+d)}+\frac{£1.00}{(1+d)^2}+\ldots\ldots+\frac{£1.00}{(1+d)^T}$$

In this case, the optimal time to retire a machine is not determined by the return, d, through passive reinvestment of the proceeds of the scrap value of a *single machine* as in Eq (9.13). The comparison in Eq (9.15) is the net income (adjusting for the rate of depreciation) of the existing machine and the right-hand side of Eq (9.15) which is the net income averaged over the life of the *next machine*, less the cost of a new machine, adjusting for the scrap value of the current machine.

A complicated analysis perhaps, but undertaken, at least implicitly, by a taxi owner-driver who mentally calculates when to replace an ageing cab with a new vehicle.

Final comment

In Chapter 4 we established the basic financial modelling that is used to calculate asset PVs. This is the core of real estate valuation. The fundamental investment Decision Rule is:

Accept a project if the Present Value (PV) $\geq$ £0.00

Reject a project if Present Value (PV) $<$ £0.00

In Chapter 5 we explored the valuation of Real Options. In decisions involving real options – a very common situation in property markets – the simple Decision Rule requires modification because a real option can have financial value. This chapter has illustrated some of the problems and insights posed by PV maximisation and value optimisation in the context of durable, exhaustible and renewable assets and resources. The applications are diverse and extensive, from decisions to demolish or refurbish a building to the management of a fleet of cars and natural resource optimisation strategies. Precise definition of the problem is critical since the solution is likely to be sensitive to the underlying model.

The choice of a discount rate, *d*, is critical in the calculation whenever the purpose is to optimise value over a period of time. A high discount rate privileges early income at the expense of later income while penalising projects or technologies with high initial set-up costs. In the case of a new building, we might expect a developer in a high real interest rate market to use cheaper materials and more rapid construction techniques to accelerate development profits and rental income, even if the longer term impact is higher maintenance costs, an increased rate of depreciation and reduced life expectancy.

In high real interest rate situations, investments with long payoff periods may not proceed at all. A forestry manager in a high discount rate market is likely to select fast-growing trees even if the timber is less durable and sells at a lower price than a slower growing variety. Even for fast-growing trees, the optimal harvest date is likely to be brought forward. In the case of durable assets, maintenance work (which will reduce profits up front) may be delayed and the life cycle of the asset abbreviated. Mining operators may choose high-grade extraction options, while mine life will be reduced. Mine operators threatened with early closure for environmental or other regulatory reasons will make similar choices. A managed fishery is likely to see a higher rate of exploitation when the discount rate rises, accepting a reduced long-term sustainable fish stock as a trade-off. It may even be economically rational to exploit the fishery resource to extinction when at a lower discount rate, a sustainable strategy for the resource may be financially optimal.

In summary, there are seldom easy or "quick-fix" solutions to the challenges posed by the limits imposed on us by planet Earth – as our 19th-century analyst predecessors fully understood. And understanding is a prerequisite for analysis.

Appendix 9A – Derivation of optimal production rates (q*) in Figure 9.2

Figure 9.1 defines a production technology, where

$$\text{Total cost } (TC) = k + aq + bq^2 + cq^3$$

Specifically in this case, we have assumed that fixed cost, $k = 0.0$ and $a = 15$, $b = -1.2$, $c = 0.1$

$$\text{Average cost } (AC) = TC / q = a + bq + cq^2$$

So, if the extraction rate $q^* = 8$ tonnes p.a., AC = £11.80 per tonne.

The income for the enterprise is defined as

$$\text{Total revenue } (TR) = zq$$

where $z = 21$

Appendix 9B – Derivation of optimal fixed rate of production, q*, in Table 9.1

At a fixed production rate q^* annual profit is

$$\Pi(q^*) = \text{TR}(q^*) - \text{TC}(q^*)$$

The problem is more readily solved in continuous, not discrete time. Maximise

$$\text{PV} = \int_0^T e^{-dt}\, \Pi(q)\, dt \tag{9.1B}$$

Noting that the *PV* of a one-dollar income stream for T years where discount rate = *d* is:

$$PV = \int_0^T e^{-dt}\, dt = (1 - e^{-dt}) / d$$

Integrating Eq (9.1B) it follows that

$$PV = \int (q)[(1 - e^{(-dT)}) / d]$$

The *PV* maximising rate of production, q^*, is established by differentiating Eq (9.1B) with respect to q and setting it equal to zero:

$$PV'(q) = \{\Pi'(q)[(1 - e^{(-dT)}) / d]\} - \{\Pi(q)[e^{(-dT)}.T / q]\} = 0$$

where $T = 120 / q^*$.

Equation (9.2B) establishes the optimising condition that for the *PV*-maximising rate of production, q^*, and a discount rate, *d*, the first term on the right-hand side,

the *PV* of the marginal profit $\pi'(q)$ discounted over *T* years

is equal to

The capitalised value of the average profit, $\pi(q)/q$, at time *T*.

This corresponds with the intuition that for $d = 0$, the *PV*-maximising rate of output at 6 tonnes per annum, where average profit = marginal profit and average profit is a maximum.

For alternative values of discount rates, *d*, the Excel Solver function establishes the fixed optimal production rate, q^* as in Table 9.1.

In the case of Table 9.2, the rate of production, $q^*(t)$, is variable. Hence *T* is not constrained, and the binding constraint is the size of the ore resource:

$$\sum_{t=1}^{T} Q(t) = 120$$

The Excel Solver function similarly solves for the maximum *PV* for alternative values of discount rate *d* (see Table 9.2).

Appendix 9C – Terminal time calculation for a chain of machines

To establish Eq (9.7)

$$PV = \sum_{k=1}^{\infty} PV_k = \frac{\int_0^T e^{-dt}\, \Pi(q^*,t)\, dt - Io + S(T)e^{-dT}}{1 - e^{-dT}} \qquad 9.1C$$

We note that from Appendix 5A, if we define *S* as the sum of the infinite expression

$$S = w + wk + wk^2 + wk^3 + \dots \; + \infty$$

Then we derive

$$S = w/(1 - k)$$

Then by analogy, if

$$S = 1 + e^{-dT} + e^{-d2T} + \cdots$$

we can express this as

$$S = 1/(1 - e^{-dT})$$

To establish the optimal time *T* to retire each of these (identical) machines we differentiate Eq 9.1C with respect to *T* and arranging terms we establish that:

$$\Pi(q^*,T) + S'(T)) = (1/w)\left[\int_0^T e^{-dt}\, \Pi(q^*,t)\, dt - I_0 + S(T)\right]$$

where

$$w = (1 - e^{-dT})/d$$

is the *PV* of a one-dollar income stream discounted at rate *d* for *T* years.

Notes

1 *Oh, I have slipped the surly bonds of earth,*
And danced the skies on laughter-silvered wings;
Sunward I've climbed and joined the tumbling mirth
Of sun-split clouds – and done a hundred things

You have not dreamed of – wheeled and soared and swung
High in the sunlit silence...
Put out my hand, and touched the face of God.
John Gillespie Magee Jr. (1922-41), High Flight, 1941
[Quoted by President Ronald Reagan in a speech following the Challenger disaster, 28 January 1986.]

2 See Chapter 6, the history of Easter Island.
3 Mill, J. S., *Principles Of Political Economy Book IV* (England: Penguin, 1970), p. 111.
4 Mill (1970), p. 87.
5 Ricardo, D., *Principles of Political Economy and Taxation* (London: John Murray, 1817), Chapter 2.
6 Ricardo (1817), Chapter 2. This chapter is also the source of the famous statement: *Corn is not high because a rent is paid, but a rent is paid because corn is high* – a key principle in land valuation. https://www.marxists.org/reference/subject/economics/ricardo/tax/ch02.htm. Accessed 27 November 2021. See also further discussion in Chapter 5.
7 Mill (1970), p. 112.
8 Marx, K., and Engels, F., c.1845. (reprint of 1939 ed.). *The German Ideology* (Amherst: Prometheus Books, 1999).
9 Keynes, J. M., Economic Possibilities for Our Grandchildren, in *John Maynard Keynes, Essays in Persuasion, 1963* (New York: W.W.Norton & Co, 1930), pp. 358–373.
10 Keynes (1930).
11 Keynes (1930).
12 Keynes, J. M., *General Theory of Employment, Interest and Money* (New York: Macmillan & Co, 1964), p. 383.
13 See Chapter 4.
14 I conducted an informal survey of local bookshops and my own personal library. It would be invidious to reference specific principles of microeconomic sources but not hard to find examples of popular microeconomics texts that shift seamlessly (but ambiguously) between flow and stock concepts in their introductions to supply/demand market models.
15 See Appendix 10A for the derivation of these values.
16 The assumption of diminishing returns establishes the second-order condition for a maximum, $\prod''(q^*) < 0$.
17 See Chapter 4.
18 This expression is derived in Appendix 10B.
19 See for example: Kamien, M., and Schwartz, N. L., *Dynamic Optimization: The Calculus of Variations and Optimal Control Theory in Economics and Management* (Amsterdam: North Holland, 1981).
20 The Excel Solver solution in Table 10.2 is derived as in Table 10.1 but with the constraint $T = 120/q^*$ from Case 1 replaced by $\sum q(t) = 120$.
21 The second-order condition for a maximum is that the sum of marginal costs rises with the rate of output.
22 The second-order condition establishes that the PV less depreciation decreases more rapidly than the investment return, $rS(T)$.
23 For discussion of this problem see, for example, Henderson and Quandt (1971), pp. 326–329.
24 See Appendix 10C for this calculation.

Bibliography

Henderson, J. M., and Quandt, R. E. 1971. *Microeconomic Theory: A Mathematical Approach.* New York: McGraw Hill.

Kamien, M., and Schwartz, N. L. 1981. *Dynamic Optimization: The Calculus of Variations and Optimal Control Theory in Economics and Management.* Amsterdam: North Holland.

Keynes, J. M. 1930. Economic Possibilities For Our Grandchildren, in *John Maynard Keynes, Essays in Persuasion, 1963,* New York: W.W.Norton & Co, pp. 358–373.
Keynes, J. M. 1964. *General Theory of Employment, Interest and Money.* New York: Macmillan & Co.
Marx, K., & Engels, F., c.1845. 1999 (reprint of 1939 ed.). *The German Ideology.* Amherst: Prometheus Books.
Mill, J. S. 1970. *Principles Of Political Economy Book IV.* England: Penguin.
Ricardo, D. 1817. *Principles of Political Economy and Taxation.* London: John Murray.

10 Investment portfolios: where does real estate fit in?

Putting it all together

Economics has been described as the "science of choice." Scarcity and choice are two sides of the same coin. The essence of scarcity is that more of one good thing necessarily means less of another. This trade-off is the cornerstone of portfolio construction. Previous chapters have looked at the models and concepts that underpin decisions (or "choices") applied to single assets and individual projects. Portfolio analysis, however, requires analysis of multiple investment opportunities. Here we revisit and consolidate *three concepts* that were discussed in earlier Chapters.

First, in Chapter 2 we stated:

> **MM Proposition I: *The market value of any firm is independent to its capital structure and is given by capitalizing its expected return at the rate, r_V, appropriate to its risk class.*[1]**

We shall now revisit the concept of "risk class."

Second, in Chapter 4 we defined the *rate of* return as

$$r = \frac{C}{PV}$$

The market rate of return, *r*, can be derived from market transaction evidence if income (*C*) and market value (*PV*) are known. In the real estate context, the rate of return is usually described as the *yield*. Importantly, we emphasised the stringent assumptions embedded in the *rate of return* or *capitalisation rate* formula. We also introduced the concept of growth (which always needs to be carefully defined as either *nominal* or *real*) and the discount rate (or expected return). We explained in Chapter 4 that in a world with zero inflation and zero growth and a perpetual income stream (and *only* in this world) the *yield (r)*, the *internal rate of return (IRR)* and the *discount rate (d)* are identical.

We shall now revisit the analysis that underpins the "rate of return."

DOI: 10.1201/9781003111931-10

Third, Chapter 5 introduced the concept of portfolios and *risk neutrality*. We stated:

> **Another way of looking at risk-neutrality is as follows: Investors are assumed to be risk averse. This does not mean that they prefer to avoid volatile assets; investors accept risk, but at a clearly defined price. For investors, the *risk* of an asset is defined not as the volatility of the asset, *but by the contribution of that asset to the volatility of a portfolio comprising all available assets.***

The underlying proposition here is that investors are not seeking the highest return for an *individual asset*; nor even the highest return for a *portfolio of assets*. Investors seek to maximise wealth from a portfolio of assets, which facilitates a stable consumption stream over time.

How do these insights fit together in a real estate context?

Diversifiable and non-diversifiable risk

At this point, we shall delve into the slippery concept of *risk*.

All human activity involves risk. Some types of risk can be avoided altogether. Working from home is a sure way to avoid a congested morning motorway. Other categories of risk can be managed or reduced but not avoided. A flood might close down a coal mine; a fire might destroy a shopping centre; a hi-tech start-up company might fail to deliver a marketable product.

How do investors price these risks?

The answer is that they don't.

These are three examples of *diversifiable risks*. By holding a big portfolio comprising many coal mines or many shopping centres or stakes in many hi-tech start-ups an investor can minimise (or diversify away) the risk of one location going out of action or one company failing. But no matter how large the portfolio of coal mines or shopping centres, an investor can't eliminate the impact of, say, an economic recession reducing the price of coal, a global financial crisis, a pandemic or a rise in Value Added Tax (VAT) that erodes consumer confidence and undermines all retail spending.

These are examples of *non-diversifiable ris*ks:

> Total financial risk = Diversifiable (or non-systemic) risk + Non-diversifiable (or systemic) risk.

Risk, of course, takes many different forms – unanticipated changes to planning requirements, zoning rules and environmental regulations can all have a positive or negative impact on the investment performance of individual assets and on whole classes of assets. But most of these risks can be captured in one of these two broad risk categories.

High systemic risks should command high returns to investors in compensation. But since asset specific or diversifiable (sometimes described as *non-systemic* or *idiosyncratic*) risks can be reduced (even eliminated) by actions taken by investors themselves, there is no reason why investors should be rewarded for this category of risk associated with these assets. In transparent and liquid markets coal mines, shopping centres and hi-tech start-ups are priced according to the non-diversifiable risks associated with those assets. Investors who choose to accept asset-specific risks are not (in theory) rewarded for these risks. This is what MM Proposition I means by "risk class."

Politicians, of course, deal every day with both categories of risks – diversifiable and non-diversifiable. Promises extracted from desperate candidates with short time horizons at the height of the electoral hustings to close down coal mines, abrogate international treaties, and build new hospitals and transport routes, are all efficiently internalised in real estate asset prices, discounted by the probability that these election commitments will ever be implemented.

The portfolio risk-return trade-off

Figure 10.1 shows two entirely hypothetical (and extreme) investment scenarios.

In **Case 1** the rates of return on two investments, A and B, are perfectly *negatively* correlated – the correlation, $R^2 = -1.0$. As the return of Investment A rises and falls over time the return of Investment B moves in exactly the opposite direction. The average return on A is 8.0 percent per annum and B delivers an average return of 4.0 percent per annum. Investments A and B are equally volatile as measured by the Standard Deviation (SD = 0.69).

In **Case 2** the returns to assets A and C are perfectly *positively* correlated – the correlation, $R^2 = +1.0$.

Now consider a portfolio in Case 1 comprising a 50 percent weighting of A and a 50 percent weighting of B. Intuition (and mathematics) shows that this portfolio will deliver an average annual return (r_p) of 6.0 percent per annum:

$$r_p = 0.50\,(.08) + 0.50\,(.04) = 6.0\% \qquad (10.1)$$

The identical calculation applies to Case 2.

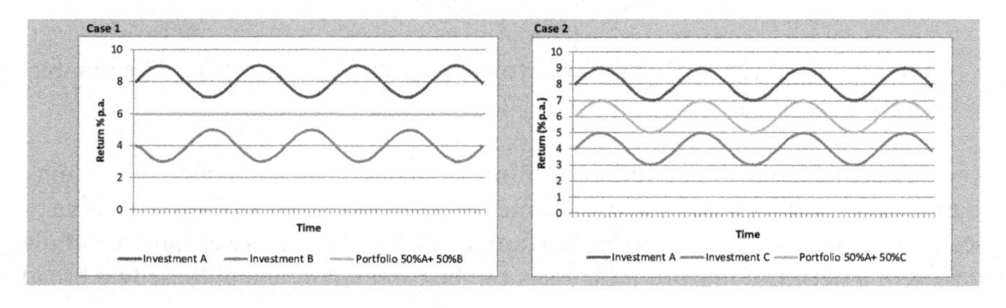

Figure 10.1 Two Investments and Portfolio Performance

In both **Case 1** and **Case 2** in Figure 10.1 the average return for Portfolio A+C is 6 percent per annum.

Principle 1: The return on a portfolio of assets is the weighted average return of the assets in the portfolio, weighted by the value of each investment held in the portfolio.

However, the volatility of the 50:50 portfolios in Case 1 (A+B) is very different from the volatility in Case 2 (A+C).

In Case 1 the volatility of the portfolio (A+B) is zero (SD = 0.0). But the volatility for the 50:50 weighted A+C portfolio is identical to the volatility of A and C individually (SD = 0.69).

From this we can derive a second principle:

Principle 2: The volatility of a portfolio depends not on the weighted average volatility of individual assets, but on the inter-temporal relationship of the returns (or covariance) of all assets held in the portfolio.

The two portfolios from Figure 10.1 are mapped on Figure 10.2 in risk-return space. Investments A, B and C all show the same volatility (SD = 0.69). So, too, does the portfolio (50% A + 50% C). But the portfolio (50% A + 50% B) shows zero volatility (SD = 0.0). Both portfolios show a weighted average return of 6 percent per annum.

Question: Which portfolio would an investor prefer, 50% A+50% B or 50% A+50% C?

Answer: Portfolio (A+B) must be preferred to (A+C); both portfolios offer an average return of 6 percent per annum; but (A+B) offers less (in fact, zero) volatility or risk. In the case of portfolio (A+C) the volatility (SD = 0.69) is identical to A, B and C individually. Portfolio (A+C) in other words offers no risk mitigation. Portfolio (A+B) offers perfect risk mitigation.

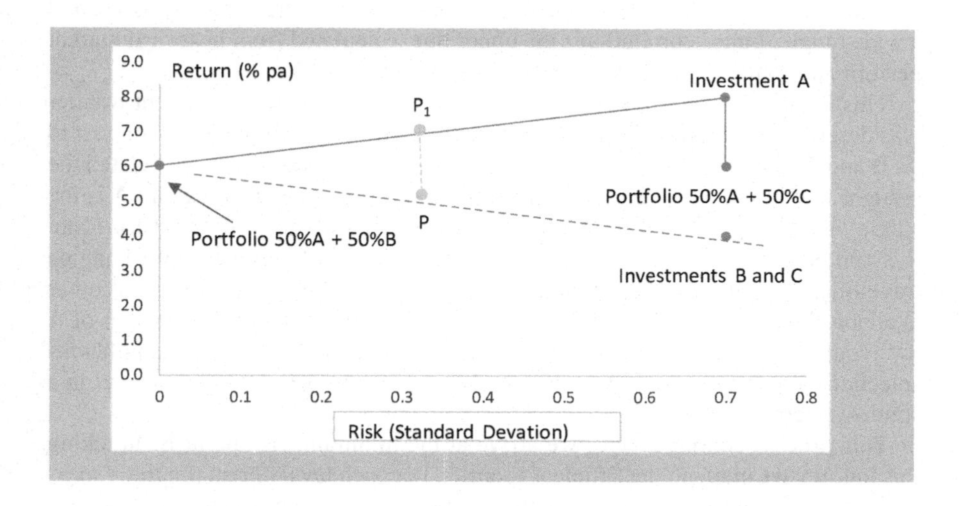

Figure 10.2 Mapping a Portfolio in Risk-Return Space

Consider now an investor who holds a 100 percent stake in Investment A. The investor gradually sells A and buys B. In Figure 10.2 this investor travels down the solid line from A, past P_1, eventually arriving at 50% A+50% B. Continuing to sell A and buy B would move the investor along the lower dotted line past P towards point B and C. Clearly any move beyond 50% A+ 50% B would be irrational. Any point on the lower line, say *P*, would have the same risk but a lower return, than point P_1 on the upper line. P_1 *dominates P*, and all portfolio options on the upper line are superior to all options on the lower line.

We can therefore disregard the lower line in making portfolio selections.

The principle: Investments B and C have identical returns (4.0 percent per annum) and identical volatility (SD = 0.69). But they behave very differently from a portfolio perspective. B has the valuable attribute that it reduces portfolio volatility, albeit at the expense of reducing portfolio return. C, on the other hand, reduces return and makes no contribution to volatility reduction.

Question: Which point on the upper line should the investor select?

Answer: We cannot definitively answer this question. Selection of the optimal point on the upper line is the task of the investor or the investor's financial adviser. An aggressive investor would choose portfolio A, which offers the highest available return, albeit with relatively high risk. A less aggressive investor would choose a point somewhere along the upper line. A risk-averse investor would be attracted to the (50% A + 50% B) portfolio, where risk is zero.

To re-state, the risk we are talking about is the volatility in the value of the portfolio, which in turn defines the wealth, and therefore the consumption pattern of our hypothetical investor. It is this consumption stream that the (hypothetical) investor cares about.

Investments A, B and C are clearly extreme cases, based on correlations of +1.0 or –1.0. The real world is a good deal less precise. In the world of many assets and a wide range of inter-correlations, the upper line, calculated from historical market performance data, typically looks more like Figure 10.3:

This is the *Efficient Frontier*. The mathematics for defining an efficient frontier was described by Markowitz (1959).[2] In a world where investments (in contrast to A, B and C in Figure 10.1) are not perfectly positively or negatively correlated the straight line in Figure 10.2 usually transforms into a smooth curve. Point A is the single asset category with the highest return (regardless of portfolio risk). From this 100 percent exposure point (equivalent to point A in Figure 10.1) we imagine travelling along the frontier by selling A and buying some combination of other available investments. Since A is the highest-return investment, divestment of A in favour of some alternative investment must reduce the return. But the judicious selection of alternative investments can also reduce the risk. Hence we travel in a south-westerly direction.

This process continues until we arrive at the minimum risk point B. In taking this journey we map out the Efficient Frontier. For each level of risk (on the *X*-axis) we select that combination of investments that delivers the highest possible return (on the *Y*-axis). Note that the Efficient Frontier in Figure 10.3 does not intersect the vertical *Y*-axis (as we did in Figure 10.2) because in practice there are unlikely to

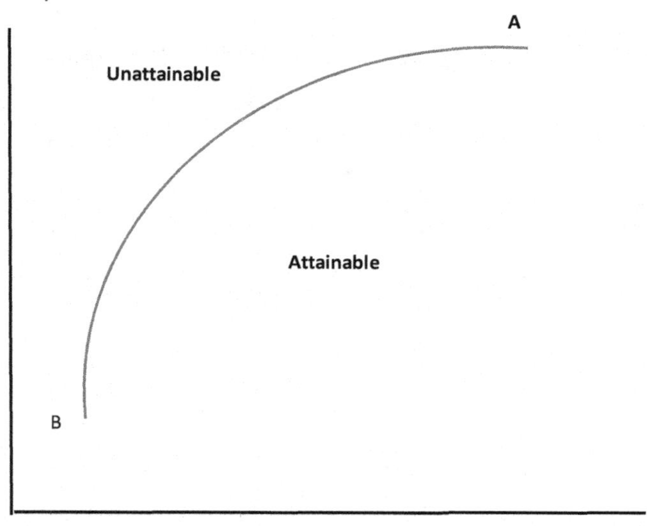

Figure 10.3 The Efficient Frontier

be perfectly negatively correlated investments (in contrast to our imaginary world in Figure 10.2).

We can of course select a portfolio at a point below (to the southeast of) the efficient frontier. This area is *attainable* (but of course, sub-optimal in the Markowitz sense). We cannot get above the efficient frontier – this area is *unattainable* (although as we shall see below, a cunning plan enables us to get around this constraint and locate our portfolio in the "unattainable" zone).

An office market portfolio

How does this analysis apply in the real world? Figure 10.4 shows the performance in risk/return space of office assets based on quarterly data (supplied by MSCI) for fourteen major European office markets (2002–21).

Over this period Stockholm was the office market with the highest average annual return (11.5 percent per annum). At the other end of the scale is Frankfurt (6.7 percent per annum). Munich was the lowest volatility market, with Dublin at the other end of the volatility scale. Obviously, these rankings are specific to the time scale over which we are measuring performance. And past performance is no guarantee of future performance. We must also be cautious in accepting the risk (or volatility) estimates shown in Figure 10.4 – a topic we consider further below.

The portfolio (P) illustrated shows the combination of markets with the highest (historical) performance given a defined level of volatility (SD = 0.045). Figure 10.4 also illustrates part of the efficient frontier derived from these data points. Stockholm is the starting point of the frontier (the market with the highest level of

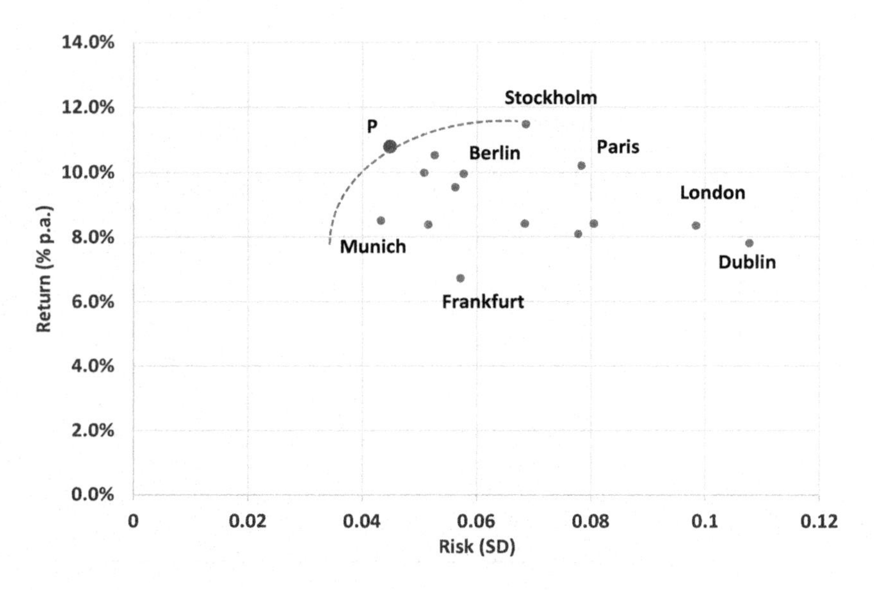

Figure 10.4 European Office Market Performance (2002–2021)

Source: MSCI.

return regardless of volatility). As we travel along the frontier (illustrated by the dotted line) toward P we divest the Stockholm market and acquire a selection of other office markets.

This specific portfolio, P, in Figure 10.4 in fact comprises the following weightings:

Amsterdam 18.8 percent
Brussels 4.3 percent
Paris 1.4 percent
Prague 33.8 percent
Stockholm 44.8 percent
Total 100 percent

Would you choose to hold this portfolio? Perhaps not; nevertheless this is the exercise that investors are, at least conceptually, engaged in when selecting a portfolio based on anticipated future rather than, as illustrated here, historical performance.[3] Figure 10.4 does illustrate that, at least in principle, it is possible to construct an office market portfolio with average annual returns not much lower than the best performing market (in this example, Stockholm) and risk (or volatility) not much higher than the least volatile market (in this example, Munich). To reiterate, this analysis is of course based on historical, not forecast performance.

As a general statement, portfolios on the efficient frontier comprise relatively few markets, and of course, the weightings defined by the Markowitz algorithm take no account of the relative capital value and liquidity of the markets – good reasons why efficient frontier portfolios should be regarded as informative, not prescriptive, as is also the case with the output of statistical forecasting models, as argued in Chapter 11. And the performance data presented in Figure 10.4 is derived from *appraised* values, not market *transactions* – some of the implications of the appraisal process are discussed in more detail below.

Asset pricing, the security market line (SML) and beta (β)

A broad-based portfolio (in the limit, a theoretical portfolio comprising all available investable assets) weighted by the total value of each asset, would deliver a return equal to the weighted average return of all these assets – let's call this the market return, r_m. An investor who holds this portfolio can expect to receive an annual return of r_m. But how do the individual assets within this portfolio perform, and how are they priced?

The SML defines this relationship (Figure 10.5).[4]

The SML establishes the formula for the Capital Asset Pricing Model (CAPM):

$$r_z = r_f + \beta\,(r_m - r_f) \tag{10.2}$$

where:

r_z represents the return required by an investor in a particular asset, z;
r_f represents a risk-free interest rate (sometimes proxied by government long-term bond rates);
r_m represents the return required by investors who select a broad asset-weighted market portfolio (in the limit, a portfolio reflecting all available investable assets), and
β represents the volatility of the asset, or asset class, in comparison with the overall market. β is a measure of systemic (non-diversifiable) risk.

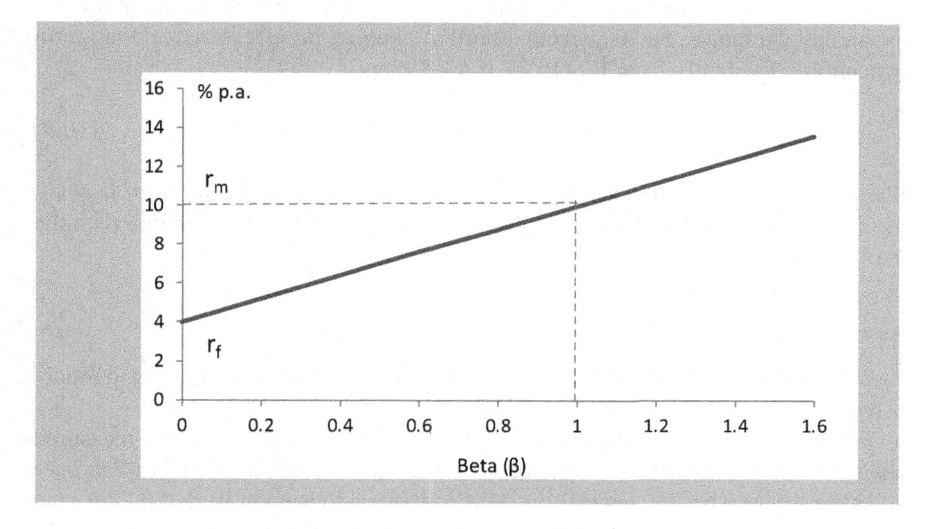

Figure 10.5 The Security Market Line

Principle: The insight of the Capital Asset Pricing Model (CAPM) is that the returns to individual assets are determined by the performance of each asset, or asset class, in relation to the overall market portfolio, r_m. Recall the discussion of systemic and non-systemic risk in Chapter 5. Assets are valued not according to their specific volatility (or risk) but according to their volatility in relation to the overall market portfolio of assets. It is the contribution of each asset to the underlying wealth of the investor that counts. The investor calls the shots. The investor's objective is to achieve stable wealth and therefore a stable consumption path over time. The investor is not a punter at a horse race – although some financial advisers and brokers may like to think otherwise.

> ...the investor cares about the volatility of consumption. [The investor] does not care about the volatility of individual assets or of [the]...portfolio if he can keep a steady consumption.[5]

Beta (β) and the equity risk premium (ERP)

Statistically, for a particular asset *z*, beta, β, is defined as:

$$\beta z = \frac{\text{Covariance}\ (r_z, r_m)}{\text{Variance}_{r_m}} \tag{10.3}$$

Since r_f represents the return available to investors who accept no risk, that is $\beta = 0$, it follows that $(r_m - r_f)$ is the *additional* return required by investors in the broader market. For investors in the equity market, this additional required return is called the Equity Risk Premium (ERP). It represents the additional return that investors in the share market require for accepting the additional risk inherent in investing in shares.

The actual value of the ERP is elusive, the subject of much debate and ever-expanding literature; but 6 percent is often taken to be a reasonable long-term benchmark. Evidently from Eq (10.2), if $\beta = 1$ then:

$$r_z = r_f + r_m - r_f = r_m \tag{10.4}$$

If $r_f = 4.0\%$ and $\beta_z = 1.0$ for a particular asset, *z*, then the required return is 4% + 6% (the ERP) = 10.0%. In other words, this asset performs strictly in line with the market portfolio.

Asset pricing and the A-REIT sector – an illustration

How does the CAPM equation (Eq 10.2) apply to real estate? Consider publicly listed Real Estate Investment Trusts (REITs).

Figure 10.6 shows estimates of beta for the FTSE 350 REIT index measured against the FTSE 350 index. Estimated over the period January 2017 to February 2023 the beta is 0.61. In comparison, MSCI reports a long-term beta of 0.78 for the MSCI US REIT Inde.[6] The rolling 12-month average beta for the FTSE 350 REIT

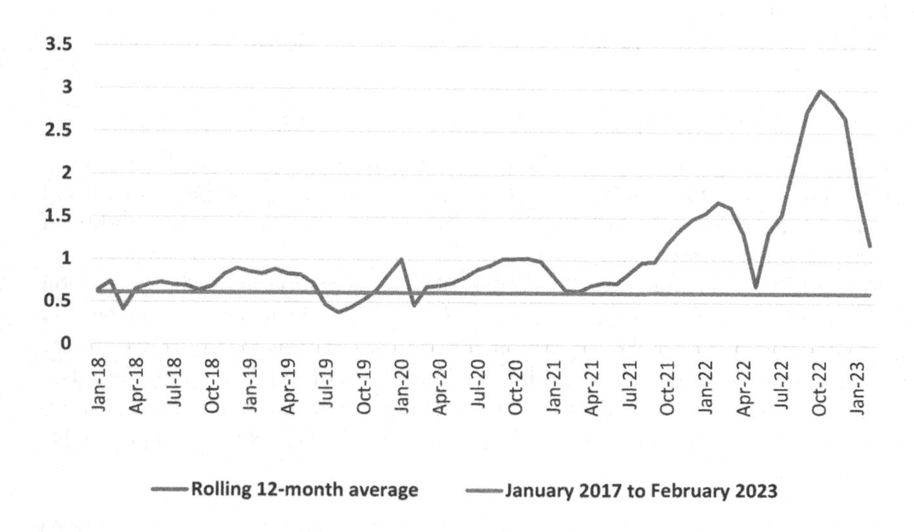

Figure 10.6 The FTSE 350 REIT Beta

Source: Investing.com.

index in Figure 10.6 shows values predominantly between 0.5 and 1.0, with a sharp spike post-2021. This spike reflects the impact of central bank interest rate tightening policies in 2021 and 2022. The REIT sector is considered to be less volatile than the overall stock market (hence $\beta < 1.0$) but it is very interest-rate sensitive. Both characteristics are revealed in Figure 10.6 – relatively stable (2017–2020) and a sharp response to rising interest rates (2021–2023).

Two further qualifications are necessary:

First, because REITs, in contrast to other listed securities, are obliged to distribute a high proportion of taxable income to shareholders, the impact of dividends on index performance is significant, and the tax treatment of REITs also differentiates them from other listed stocks. Therefore a beta calculated purely from capital growth, which is the basis for the analysis in Figure 10.6, rather than total returns (capital growth plus income yield) is potentially misleading.

A range of beta estimates and measurement methodologies are available in the market. Generally, however, analysis of public REIT performance in the UK and other markets supports the proposition that REITs are generally low volatility, or defensive, assets. A detailed analysis of alternative methodologies for calculating betas has identified substantial differences in methodology between commercial data providers – but confirmed the evidence that REITs are, in general, low beta investments – less than 0.4 in most cases.[7]

Let's take the long-term average value of beta, say, 0.4. The current (February 2023) yield on 30-year UK gilts is 3.58 percent, and we adopt the convention that this represents the risk-free rate, r_f.

Then from Eq (10.2) the implied (nominal) return to the UK public REIT sector in February 2023 was

$$r_z = 3.58 + 0.40\ (6.0) \qquad (10.5)$$
$$= 5.98\%\ \text{p.a.}$$

Note that this is significantly below the required return from the overall share market, which is r_f + ERP = 3.58% + 6.0% =9.58%. Investors are apparently willing to accept lower returns from REITs as a trade-off against the relative stability of returns provided by the sector.

Second, the beta illustrated in Figure 10.6 is an *equity beta.* The discussion around the SML and in Figure 10.5 is actually built around *asset betas* rather than *equity betas* as in Eq (10.5). As demonstrated in Chapter 2, higher levels of debt imply increased volatility and therefore a higher level of required equity returns – but the asset beta is independent of the debt/equity ratio. Implicitly the return derived in Eq (10.5) is for all-equity investments. A more comprehensive equation for the asset beta, β_A, is:

$$\beta_A = \beta_E(E/V) + \beta_D\ (D/V) \qquad (10.6)$$

where:

β_E is the equity beta
β_D is the debt beta, likely to be a low number (even $\beta_D = 0$ in the case of bank debt if the rate of interest is fixed for the duration of the loan)
V is $D + E$, the value of the enterprise.

Beyond the efficient frontier

At this point, we combine and apply the insights from Figures 10.3 and 10.5 (Figure 10.7).

The efficient frontier (Figure 10.3) was implicitly constructed from market transaction data. All assets, therefore, reflected some volatility. In all cases SD > 0.0. We now re-introduce an additional asset. This is a risk-free asset, sometimes proxied by government sovereign bonds. An immediate qualification: if these are *nominal* bonds (as we employed in Eq 10.5) they are exposed to changes in the inflation rate. Real (inflation-indexed) bonds are a better proxy when seeking a "risk-free" asset.

However, if we can assume the existence of a risk-free asset then it must be represented by some point, r_f on the vertical axis. And recalling Figure 10.2 we can construct a portfolio comprising a proportion *x%* the risk-free asset and *(1–x%)* of the market portfolio, r_m. on the efficient frontier. Clearly any point on the straight line, for example, *P*, will dominate any point on the efficient frontier except r_m where the lines coincide. And since all investors (in theory) will make the decision to locate their portfolios along the straight line r_fPr_m it follows that the market

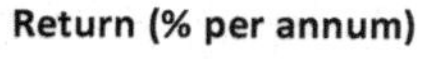

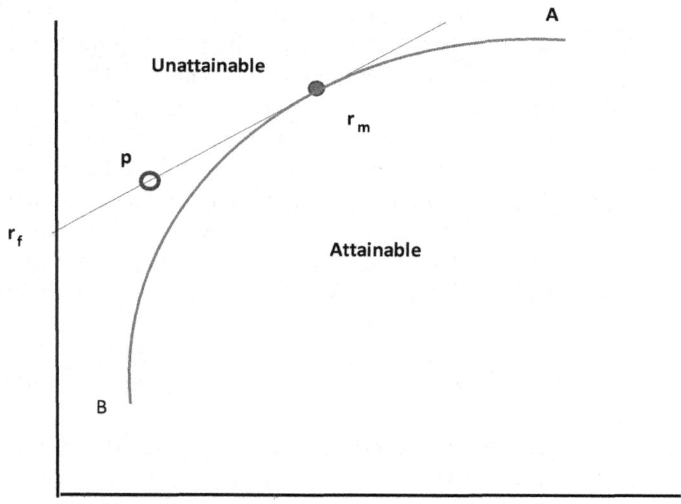

Risk (Standard Devation)

Figure 10.7 The Capital Market Line (CML): Crossing the Frontier

portfolio r_m is the only market portfolio that can exist. All investor portfolios will comprise a proportion of the risk-free asset and the market portfolio (comprising all other assets). The precise proportions will vary because investors will locate themselves at different points along r_fPr_m, depending on their individual risk preferences. This is called the Capital Market Line (CML).

Appraisal-based indices and de-smoothing – apples and oranges

In a previous section we explained that the required return for any asset (including real estate) is given by Eq (10.2), the Capital Asset Pricing Model. Investment in publicly listed REITs conforms to the same model, and because market evidence shows that the REIT equity beta is typically < 1.0, it follows that the required return on a portfolio of REITs is less than the required stock market return.

There is, however, a complication. Share market and bond market indices, including REIT indices, typically reflect actual market transactions. In the case of direct real estate, performance indices are often based on valuations (or "appraisals") rather than transactions. This is because transaction evidence may be sparse, out-of-date or unavailable. In addition, real estate assets are heterogeneous, so even recent transaction data may reflect characteristics unique to specific assets and therefore not representative of trends in the broader market.

The FTSE 350 REIT data in Figure 10.6 is derived from actual market transaction data. In contrast, Figure 10.4 shows the risk/return characteristics of a selection of office markets, based on MSCI index data. This is an appraisal-based index, not a market-traded index such as the FT 350 REIT and other stock market indices. Comparisons of asset classes (equities, bonds, REITs, cash, real estate) typically use market performance indices from a range of sources – S&P, MSCI, NCREIF, for example. The methodologies used to derive these indices may differ.

Comparison of performance between asset classes (say, direct real estate and equities) requires caution if the data is derived from different processes; the risk is an apples vs oranges comparison. Appraisal-based indices often show a tendency to smooth out market volatility as valuers and analysts often base their assessments on long-term trends rather than respond immediately to only the most recent, and possibly unrepresentative, transaction evidence. If this is how valuations are constructed, real estate valuations presented for financial reporting purposes only partially adjust previous valuations in light of recent evidence.[8] The impact of a smoothing process is that the volatility of real estate is understated in risk/return comparisons with equity or bond investments.

Various adjustments have been proposed to reverse, or "de-smooth" the impact of the appraisal process. First- or second-order autoregressive models or the direct application of unsmoothing factors have been proposed. For example, the outcome is a lagged adjustment valuation process which might be expressed as:

$$V_t = k\,M_t + (1 - k)\,V_{t-1} \tag{10.7}$$

where:

V_t represents appraised value at time t
M_t represents the market value (established from transactions evidence) at time t
V_{t-1} represents appraised value at time $t-1$
k represents the adjustment factor ($0 \leq k \leq 1$)

If $k = 0.1$ for example, the current appraised value will apply a 10 percent weighting to the most recent transaction evidence and a (1.0 − 10% = 90%) weighting to accumulated long-term evidence. When valuations are combined into market indices, the impact of the valuers' adjustment process on measured market performance is to introduce smoothing (the quarter-to-quarter or year-to-year changes are understated) and time lags (market turning points are delayed).

A range of alternative techniques is suggested to estimate the adjustment parameter, *k,* including regression analysis. As a broad statement the adjustment parameter, *k,* is often adopted in the 0.4–0.6 range. Giliberto (1992) reported on a US survey which suggests that the true volatility of non-residential real estate is around half that of equities. While this may be a convenient rule-of-thumb, analysis of actual appraisal-based office market data however suggests that the value of *k* varies significantly between real estate markets and also through time, perhaps reflecting changes in market liquidity and the impact of market cycles on the appraisal process.

Notes

1 Modigliani and Miller (1958).
2 The mathematics for deriving the efficient frontier is complex. However, the Microsoft Excel Solver tool can be employed to derive a frontier.
3 It is likely that in practice a quantitative exercise would impose constraints on the selection process, for example, maximum and minimum portfolio weightings: London >=10%; Warsaw <= 4%. In addition, anticipated future performance is relevant rather than past performance. The portfolio illustrated in Figure 10.4 imposed no such constraints and reflects the historical performance of these markets.
4 The SML was derived by Lintner (1965) and Sharpe (1966) following the work of Markowitz (1959).
5 Cochrane (2005), p. 14.
6 MSCI (2022).
7 Corgel and Djouganopoulos (2000).
8 The fact that valuers adjust their valuations only partially in light of new transaction evidence is not necessarily a criticism. Quan and Quigley (1991) have demonstrated that smoothed or lagged appraisals can be *optimal* in a thinly traded market.

Bibliography

Cochrane, J. H. (2005). *Asset Pricing*. Princeton, NJ: Princeton University Press.
Corgel, J. B., and Djouganopoulos, C. (2000). Equity REIT Beta Estimation. *Financial Analysts Journal*, 56(1), pp. 70–79. DOI: 10.2469/faj.v56.n1.2332.
Giliberto, M. (1992). *Real Estate Risk and Returns: 1991 Survey Results*. New York: Real Estate Research Papers, Salomon Bros.
Investing.com https://www.investing.com/indices/ftse-350-reits-historical-data (Accessed 4 Feb 2023)
Lintner, J. (1965). The Valuation of Risk Assets and the Selection of Risky Investments in Stock Portfolios and Capital Budgets. *The Review of Economics and Statistics,* 47(1), pp. 13–39.
MSCI Index factsheet. December 30 December 2022.
Markowitz, H. M. (1959). *Portfolio Selection: Efficient Diversification of Investments.* New York: John Wiley & Sons.
Modigliani, F., and Miller, M. (1958). The Cost of Capital, Corporation Finance and the Theory of investMent. *American Economic Review,* 48(3), pp. 261–297.
Quan, D., and Quigley, J. (1991). Price Formation and the Appraisal Function in Real Estate Markets. *The Journal of Real Estate Finance and Economics,* 4(2), pp. 127–146.
Sharpe, W. F. (1966). Mutual Fund Performance. *The Journal of Business,* 39(1), pp. 119–138.

11 Inside the crystal ball – A user's guide[1]

What is forecasting for?

Forecasts, explicit or implicit, under-write all decisions. Sometimes forecasts are an explicit input into a decision-making process; often however forecasts are implicit, or worse, a postscript artfully constructed to support a decision already made. Forecasting is particularly important in real estate. Capital commitments are often of long duration; asset values and transaction costs are high; information can be scarce or out-of-date; commitments can be costly and sometimes impossible to reverse.

This has two implications. First, resources allocated to the forecasting process are likely to be small relative to the costs of a transaction – especially when assessed in relation to the potential costs of a "bad" decision. Second, the forecasting process, the choice of technique and how forecasts inform decision-making merit attention in proportion to the costs and risks of the prospective investment itself.

Real estate forecasting poses specific challenges. The 2008 Global Financial Crisis (GFC) and the COVID-19 global pandemic have invigorated debates about forecasting techniques and processes. Liquidity in real estate markets is typically low in comparison with equity and bond markets. Sparse transaction evidence means analysis must often rely on partial or stale information to assess market trends and identify turning points. The fact that every physical real estate asset is unique amplifies this task. Market characteristics (such as lease duration) and popular metrics (such as definitions of rents and capitalisation rates) vary over time, between countries and often between market sectors in the same country.

Emphasising the importance of forecasting in real estate markets, decisions are often made against a backdrop of macroeconomic volatility and recent or potential regulatory changes, to which real estate assets, long-duration and immobile, are particularly vulnerable. These considerations argue for a considered commitment of resources to forecasting prior to, and through the life of, a real estate investment or management decision. The forecasting conundrum does not end with the acquisition of the asset. The characteristics of real estate markets argue for a careful examination of how forecasts are constructed and how the insights and uncertainties revealed are then integrated into the decision-making process. To make a useful contribution to the investment process

DOI: 10.1201/9781003111931-11

forecasts must be contestable, the methodology transparent, the risks identified and the predictions revisited over the life of the commitment.

Twenty-first century – A new crystal ball game?

The 2008 Global Financial Crisis (GFC) was widely interpreted as a joint failure of public regulatory oversight and private market analysis. A predictable response to the GFC was the belief that more timely, more accurate and granular data would facilitate public policy interventions and private investor responses in the future. While this is at least an arguable proposition, the COVID-19 pandemic has now demonstrated that no level of government regulation or market scrutiny can anticipate wild-card "Black Swan" events, sometimes of global significance. Arising from these crises, increased attention is now directed not only towards the accuracy of forecasts and the confidence limits attached to them. As important are reflections on how forecasts are prepared, by whom and how forecasting processes are embedded in public policies and incorporated into private investment decisions.

Examples of *public policy* responses to the GFC crisis include the joint International Monetary Fund and Financial Stability Board *G20 Data Gaps Initiative*[2] and, in the United Kingdom, the Investment Property Forum–sponsored *Vision* report published in 2014.[3] The ambitious purpose of this report was to develop a framework for the real estate debt finance market "ensuring an attractive, efficient and stable market, in the time frame of the next cycle."

In comparison, the analysis and science of forecasting processes appear to have commanded less formal attention post-2008 from *private-sector* real estate practitioners. A decade later, and as the world adjusts to living alongside COVID-19, this is a good time to review the situation.

Some headline questions

Anyone with a serious interest in forecasting (which means almost everyone in a real estate decision-making, analysis or advisory role) must address and answer a range of questions, including:

- How is it that some (admittedly, very few) forecasters seem able to consistently "beat the odds"?
- Do some forecasting processes deliver superior outcomes and under what circumstances? What, for example, is the optimal frequency for revisiting and revising forecasts?
- Is there a positive payoff from retrospective performance analysis?
- Should formal statistical modelling be bolted onto other techniques such as third-party industry surveys or expert market judgement – and if so, how?
- When is forecasting an appropriate in-house function and when is it better contracted out?
- Who should manage, who should participate in and at what stages, an in-house forecasting process?

- Do scenarios, high/low sensitivity analysis and confidence limits, as opposed to point forecasts, lead to improved decisions?
- Do the answers to any of these questions vary with the investment time horizon, the market sector or the nature of the organisation?
- Who "owns" the forecasts in-house – where does the buck stop?

These questions can be ignored, but not avoided. Answers to all these questions remain contestable, and it is unlikely that one size fits all decisions or investment formats. But tentative suggestions are emerging from published analyses of forecasting track records.

Most recent research into forecasting processes and practices is emerging from outside the real estate sector, but casual observation and some evidence suggest increased attention to forecasting practices within the real estate sector.[4] In part, this may have resulted from a migration of portfolio managers and analysts from other asset classes accompanied by increased capital allocation to the sector following the acceleration of globalisation of real estate markets post-2008. These new recruits bring insights and processes that often challenge preconceived ideas, even if their tools and models sometimes require modification for application in real estate markets.

In 2016, Philip Tetlock and Dan Gardner published *Superforecasting: the Art and Science of Prediction*,[5] which sets out to identify the characteristics and practices of a small group of successful so-called "superforecasters." While their analysis does not specifically include the real estate sector (and of course their findings have themselves been contested), many of their observations and findings are thought starters for a discussion about real estate market forecasting.

The top-down macroeconomic perspective

Many forecasting exercises start with a macroeconomic overlay. This typically includes general economic health indicators – GDP growth, unemployment, consumer spending, and capital investment.

The common experience is that both the 2008 GFC and the COVID-19 pandemic were unanticipated in timing, duration and amplitude, while the long-term consequences of both events are still emerging. While some forecasters will claim *expost* to have "called" any important market shock or turning point we must remember a broken clock is accurate twice every day. The reputations of the top-down clairvoyant economist, as well as the bottom-up seat-of-the-pants practitioner, have taken a battering. The confidence attached to forecasts delivered with authority and gravitas by invitation in an oak-empanelled boardroom has diminished. It can fairly be observed that the continuing demand for market forecasts is derived not from confidence in the quality of the product but from an insatiable thirst for information and a yearning for certainty. The market for forecasts is demand-led.

Nor is this a casual observation.

Recessions, as the IMF observes in a recent report, are not rare.[6] Economies are in recession 10 percent–12 percent of the time. Analysis by the IMF shows out of 153 recessions (1992–2014) economic forecasters correctly called a recession only five times (or 3.3 percent of the occasions) in the April of the year preceding the recession. In October of those same years (that is, three months prior to a recession year), the success rate rose sharply to 77 percent. The improvement in the forecast success rate as the time horizon contracts from nine months to three months is substantial but cold comfort to investors in long-term assets, such as real estate.

If this is the record of accomplishment of macroeconomic forecasters making short-run predictions, it seems likely that forecasts at the individual market (or asset) level, and over much longer horizons, will be less accurate. Nor is this performance limited to private sector forecasts. IMF analysis shows "... forecasts of the private sector and the official sector are virtually identical; ... both are equally good at missing recessions."[7]

As a further challenge to accepted practice, the global shift in recent years of bond yields, yield spreads and the impact of information technology on retail habits, office occupancy and inventory management calls into question previously reliable benchmarks and long-standing rules-of-thumb – workspace ratios in offices, tenancy mix benchmarks in retail malls, stock-to-sales ratios in warehouses.

Decision rules and valuation metrics that rely on reversion to the mean are open to challenge.

Given that point forecasts usually come with wide (implicit) confidence intervals that widen sharply as the time horizon extends, a practical response may be to adopt scenario or sensitivity analysis. At the least, third-party forecasts should rarely be adopted without question. As Tetlock and Gardner observe, "the list of organisations that produce or buy forecasts without bothering to check for accuracy is astonishing."[8]

Further, the 2018 IMF report suggests, one source of forecasting failure is behavioural – a smoothing process arising because forecasters are reluctant to change their position in response to new data. This proposition is analogous to the well-known analysis of "smoothing" identified in appraisal-based real estate indices.[9]

> Appraisers have to make an optimum assessment of value, based on fundamental variables and market information, including transactions and a market-wide appraisal index. However, transaction prices are a noisy signal and it is the appraiser's role to extract the signal from the noise in an efficient manner. This involves a process of optimal combination of past and current information and leads to appraisal smoothing.

Clearly, there are trade-offs between incorporating recent (but noisy and potentially erratic) data into the analysis and seeking to identify important turning points. As a practical suggestion, to stress test the tension between behavioural inertia and recency bias, greater contestability in the in-house forecasting process might lead to an improvement, or at least to an analysis that is more robust and transparent. The US Federal Reserve Board, for example, has a formal process to challenge its own internal forecasts.[10]

The bottom-up market perspective

If macroeconomic forecasters are vulnerable to the virus of adaptive expectations, bottom-up, market-level forecasters risk the opposite effect. If on-the-street leasing, sales and capital market operatives are invited to contribute to an in-house forecasting exercise, as they surely should, the possibility exists that their latest successful transaction is proposed as a key benchmark for resetting market trends.

Two practical suggestions may serve to address the well-known problem of recency bias.

- The *first* is to review forecasts at frequent and regular intervals. Frequent (say, quarterly) reviews reduce the risk that any single transaction has a disproportionate weight in an overall market assessment.
- The *second* suggestion is to schedule regular formal forecast meetings supported by a record of proceedings. The minutes of the previous meeting facilitate continuity, contestability and transparency: "So, three months ago you said *that;* now you are saying *this.* Why did you change your mind?" Record keeping is important. Busy people are unlikely to recall in any detail what they said at an internal meeting twelve weeks ago. The knowledge that you may be held to account (even in the benign setting of an open forum discussion) encourages reflection.

While these suggestions seem mundane, the structure and the formality of a paper trail provide the basis for another of the key success factors identified by Tetlock and Gardner – the retrospective debriefing:

> Forecasts aren't like lottery tickets that you buy and file away ... they are judgements that ... should be updated in light of changing information.

A less obvious payoff from a formal review process is that everyone, including in particular market sales and leasing operatives, are subject to the discipline of thinking strategically (some, perhaps, for the first time) about trends and drivers of markets where they operate: For example:

> What *are* the drivers? *Why* do I think that? How *solid* is that argument? How do you respond to this *contrary* evidence?

Repeated over time, the quality of contributions by bottom-up operatives to regular forecasting forums is likely to improve as they are exposed to challenges to their opinions and conflicting market insights. Nor is the information process a one-way street. An implicit education programme arising from the regular sharing of information by all forecasting team participants is one of the positive, if unplanned, pay-offs from a structured in-house forecasting process.

Formalising the modelling process

Real estate markets, which are notoriously cyclical, are natural candidates for statistical analysis. Statistical modelling sharpens debate. It mandates transparency – there is nowhere to hide. Regardless of the accuracy of the forecasts, the process of modelling imposes discipline by requiring explicit examination of market drivers; long-held beliefs about market leads and lags are tested; irrelevancies are subject to challenge and discarded. Imperfect knowledge and preconceived assumptions about market relationships are exposed or at least updated. Usefully, the sparse specification of a statistical model can stimulate a search for *nonquantifiable factors*, such as emerging regulatory policies and changes to infrastructure, which can provide good reason to override model forecasts that are based on a blind projection of the recent past. In addition, a formal statistical model provides a convenient platform for scenario analysis and "what if" exercises, as well as retrospective reviews of past successes and failures.

The benefits of formal statistical modelling are likely to increase in the future as more timely and better quality real estate market data become available and as historical data sets lengthen. But even a poor model has benefits. Warwick McKibbin, a Senior Fellow at the Brookings Institution observes, "… a bad model with transparent assumptions is better than arbitrary analysis based on wishful thinking."[11]

Good advice – but only if you are prepared to stress test the assumptions before and after decision time.

The bottom line – prescriptions and practical guidelines

Often forecasts are called for on an *ad hoc* basis under the pressure of a particular transaction or event. More dangerously, forecasts are called for in support of an acquisition or disposal decision already at an advanced stage towards finality.

Figure 11.1 sets out a suggested structure for thinking about an in-house real estate forecasting process. The starting point is assumed to be third-party macroeconomic inputs which provide the framework for the forecasting exercise.

The forecasting process itself can be notionally divided into three, overlapping, time frames.

In the short term (say, zero to three years) weight should be attached to bottom-up market evidence. It is likely, for example, that sales and leasing operatives can offer insights on prospective transactions, lease expiry profiles, market sentiment, tenant intentions and enquiries not yet evident in market data. Recent transactions can provide important evidence of market metrics. As important, and often overlooked, is information about transactions that *do not occur* – and therefore are more difficult for arm's length market observers to track. Aborted transactions can be as informative as successful transactions in assessing market price points and the weight of prospective capital. While a good deal of bottom-up information is qualitative and impressionistic it nevertheless fills a gap because published "hard" top-down data comes with a lag and may be biased towards high-value and successful transactions. Both sources of information constitute valid market evidence.

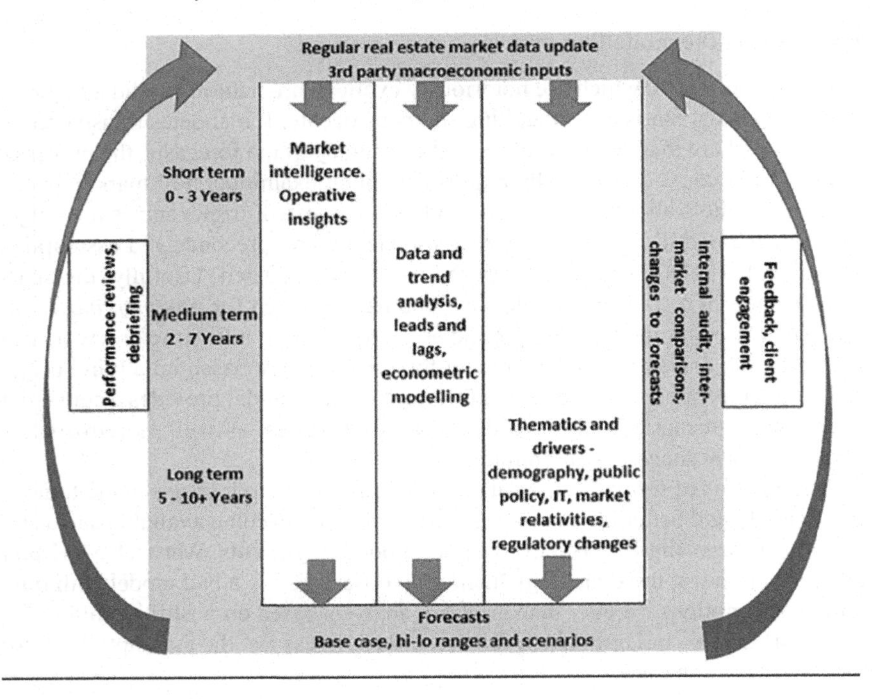

Figure 11.1 A Real Estate Forecasting Platform

The medium term (say, two to seven years) is the time frame over which statistical modelling is likely to deliver the most value. This time frame encompasses most real estate cycles and by analysing leads and lags, elasticities of demand and other drivers a statistical model can provide a framework for contestable analysis, or at least a platform for fruitful debate. Simulations, scenarios and confidence intervals can be harvested from a statistical model.

Modelling comes with several trigger warnings, however. Confronted by the differences (which may be substantial) between model forecasts and the workings of market intuition the risk is that (a) model forecasts may be taken as an authentic override of intuition or conversely (b) as clearly "wrong" because they challenge an existing consensus. Pragmatic advice is to (a) examine the model forecasts carefully – they may be trying to tell you something important, but (b) discard or amend (but keep on file) the forecasts in the face of convincing contrary information. Over time, an established track record may help to build an appropriate mix of scepticism and confidence in the modelling. Importantly, a model is a tool, not a solution, to the forecasting problem.

The long term (five to ten years+) requires an assessment of structural drivers – demography, technology, and the impact of infrastructure investment, for example. Since, as evidence shows, forecasting accuracy usually diminishes sharply as the time horizon increases, the case for scenario construction, or at least a checklist of high-level drivers and wild cards, is persuasive.

In the interest of promoting discussion, an in-house forecasting process should

- **Set the ground rules for the forecasting team**
  Define formally the responsibilities of the forecasting team and the purpose of the forecasting process. How frequently will the team meet, which sectors, markets or assets are to be the subjects of the forecasts and what variables will be forecast – rent, vacancy, yields? What are the sources of data inputs – third party, in-house? Who is the audience for the forecasts? What is the forecast time horizon? Who will manage the process – the meetings, the agenda, the minutes, the modelling? What constitutes a meeting quorum? Is there a formal decision rule in the event of disagreement – indeed, is a formal rule necessary or desirable? Is a process in place for formal retrospective reviews? Do the ground rules provide immunity from corporate politics?
- **Review predictions at regular intervals**
  Forecasts do not occur in isolation – they are normally built around a top-down macroeconomic perspective and a narrative that needs to be updated – as the data and facts change, so should forecasts, even if it's only the confidence intervals around base case numbers. Market drivers are likely to be third-party data – data on the real macro-economy and financial markets, for example – which is published on a quarterly basis – potentially a suitable interval to review the data and assumptions underlying the forecasts.
- **Involve a team of core members, including market-facing operatives**
  A nominated core team ensures consistency and should be selected to include the diversity of skills, functions and seniority. Market-facing operatives may well be the first to identify bottom-up game-changing trends or themes; at the risk of "false positives" they are a valuable source of early warning alerts. A trade-off for the chairperson to manage is between ensuring that forecasts are not shifted in response to high-profile but isolated events and the tendency of participants to defend their forecasts up to the last against new evidence. A useful agenda item is to develop and review a list of "black swan" or "wild card" events that might disrupt the base case forecasts – "what could go wrong?." By definition these events are unpredictable, but a hypothetical list may encourage a team culture of mental exploration and lateral thinking.
- **Provide a forum for contestability**
  Disagreement but not discord, dissent but not dispute: a successful forum calls for skilled chairing of the discussion. Forecast meetings are worthless if one (or two) senior team members are allowed to dominate proceedings. It is important that participants be required to present a clear point of view supported by argument and evidence, and then allowed subsequently to change a position without loss of face. One suggestion is to have a formal appointment of a team member to challenge the consensus, a "counsel for the prosecution." An argument presented in writing and in advance of the meeting – on a one-sided sheet of A4 paper, maximum – may facilitate a structured and productive debate.

- **Conduct formal retrospective evaluations of previous forecasts**
  The forecasting meeting is not a talking shop. The purpose of the meeting is to share insights and information, not to ratify a prior agenda. Members must accept responsibility for their positions – and be allowed to review or reverse positions as required. In practice, members are unlikely to recall what they said three months previously – which is why (brief, bullet-point) minutes of the meeting are required, to be circulated immediately after each meeting (to allow for challenge and review) and re-presented at the subsequent meeting as a thought starter. Since the team members (one hopes) are busy people, meetings should start and terminate on time. This is unlikely to allow for a formal critique of previous forecasts. Retrospective evaluations of previous forecasts are a separate exercise, with similar but not necessarily identical attendees. The results of these reviews should be presented, perhaps on an annual basis, to the internal customers of the forecasts.
- **Support forecasts with statistical modelling where possible**
  Statistical modelling has several advantages. It assists in identifying important drivers and relationships and challenges or eliminates others. Model forecasts provide a starting point for (not a means to close down) debate. Models also clarify the difference between statistics and personal judgment (which should be seen as another key ingredient, in no way qualitatively inferior to the output of the model). It is generally valid that successful modelling does not necessarily require statistical complexity. Given the availability and quality of data in many real estate markets, the benefits of additional statistical sophistication tend to diminish rapidly. The benefits of statistical modelling are likely to vary by market and sector – for example, historical data on a CBD office market is likely to be available and of good quality; an emerging office market may have a limited statistical track record. Data sets on retail shopping malls may be richer than data on the logistics sector although the diversity within and between retail sectors may limit detailed modelling. Residential markets often are backed up with quite detailed data on residential construction, rental vacancy rates and dwelling prices.
- **Explicitly identify key assumptions, reservations and uncertainties**
  Forecasts that profess to lift the veil concealing the future can convey a false sense of certainty. Point forecasts in the long-term future, absent an assessment of confidence limits, no explicit identification of exogenous variables or underlying assumptions – these are all warning signs.

The outcome of a successful forecasting process is more than a set of numbers, although the numbers are important. It may be that the participants in the process, and their internal or external customers, emerge with *reduced confidence* in their predictions of the future than they had at the start of the process; but with an *enhanced understanding* of the risks, opportunities and assumptions upon which they base their decisions.

Notes

1 This chapter is adapted and extended from an article by the author published in *Institutional Real Estate – Asia Pacific*, October 2019.
2 IMF/FSB G-20 *Data Gaps Initiative, Users Conference on the Financial Crisis and Information Gaps*, July 2009 https://www.imf.org/external/np/seminars/eng/2009/usersconf/index.htm and subsequent reports.
3 Investment Property Forum (2014), *A vision for real estate finance in the UK*, and Investment Property Forum (2017), *Long-term value methodologies and real estate lending.*
4 See for example *The Future of Property Forecasting*, IPF Research Program, November 2012. The empirical tests of alternative forecasting techniques in this report are suggestive, although the market under review – the City of London office market – is an unusually rich source of reliable current and historical real estate data.
5 Tetlock, P., and Gardner, D. (2015), *Superforecasting: The Art and Science of Prediction* (Random House, 2015).
6 IMF An Z., Jalles J., and Loungani, P., How Well Do Economists Forecast Recessions? IMF Working Paper, 2018, WP/18/39.
7 IMF (2018) p. 1.
8 Tetlock et al. (2015).
9 Geltner, D., MacGregor, B. D., and Schwann, G. M., Appraisal Smoothing and Price Discovery in Real Estate Markets, *Urban Studies*, 40, nos. 5–6 (2003).
10 See RBA forecasting analysis Pagan, A., and Wilcox, D., *External Review – Reserve Bank of Australia Economic Group. Forecasts and Analysis* (Report to the Reserve Bank of Australia, 2015), p. 15.
11 Source: *Australian Financial Review* 27 March 2019.

Bibliography

An, Z., Jalles, J. T., and Loungani, P. 2018. How Well Do Economists Forecast Recessions? *International Finance,* 21(2), pp. 100–121.

Armstrong, J. S., 2001. Selecting Forecasting Methods, in *Principles of Forecasting,* Springer, Boston, MA, pp. 365–386.

Geltner, D., MacGregor, B. D., and Schwann, G. M. 2003. Appraisal Smoothing and Price Discovery in Real Estate Markets. *Urban Studies,* 40(5–6), pp. 1047–1064.

IMF/FSB. July 2009. *G-20 Data Gaps Initiative, Users Conference on the Financial Crisis and Information Gaps.* Washington, DC: IMF.

Investment Property Forum. 2014. *A Vision for Real Estate Finance in the UK.* London: IPF.

Investment Property Forum. 2017. *Long-Term Value Methodologies and Real Estate Lending.* London: IPF.

IPF Research Program, November 2012. *The Future of Property Forecasting.* London: IPF.

McKibbin, W. 2019. How Should Technocrats Count True Cost of Cooling the Climate? *Australian Financial Review.*

Pagan, A. & Wilcox, D. 2016. *External Review–Reserve Bank of Australia Economic Group Forecasts and Analysis.* Reserve Bank of Australia. https://www.rba.gov.au/speeches/2016/pdf/sp-ag-2016-04-06-external-forecast-review-final-report.pdf

Rees, D. 2019. Making Forecasts: Prioritizing Processes, Prescriptions, Real Estate Investment Forecasting. *Institutional Real Estate,* 31(9), pp. 27–30.

Tashman, L. (ed) 2012. *The Forecasting Process: Guiding Principles, Process Design and Management of Change.* Medford, MA: Institute of Forecasters.

Tetlock, P., and Gardner, D. 2015. *Superforecasting: The Art and Science of Prediction.* London: Random House.

Index